TM 9-1425-386-10-1

PERSHING II WEAPON SYSTEM OPERATOR'S MANUAL

(SYSTEM DESCRIPTION) TECHNICAL MANUAL

(PERSHING II FIELD ARTILLERY MISSILE SYSTEM)

by HEADQUARTERS, DEPARTMENT OF THE ARMY

TECHNICAL MANUAL

HEADQUARTERS,
DEPARTMENT OF THE ARMY
WASHINGTON, D.C. *1 June 1986*

Operator's Manual

PERSHING II WEAPON SYSTEM
(SYSTEM DESCRIPTION)

(PERSHING II FIELD ARTILLERY MISSILE SYSTEM)

REPORTING ERRORS AND RECOMMENDING IMPROVEMENTS

You can help improve this manual. If you find any mistakes or if you know of a way to improve the procedures, please let us know. Mail your letter, DA Form 2028 (Recommended Changes to Publications and Blank Forms), or DA Form 2028-2 located in the back of this manual direct to: Commander, U.S. Army Missile Command, ATTN: AMSMI-LC-ME-PMB, Redstone Arsenal, AL 35898-5238. A reply will be furnished to you.

CHAPTER 1
INTRODUCTION

1-1. SCOPE.

a. *Type of Manual.* Introduction to the PERSHING II (PII) Field Artillery Missile System. The presentation of information is directed toward personnel who require a general knowledge of the system. This manual contains descriptions of:

- System equipment
- Missions and firing roles
- PII battalion organization
- Deployment
- Maintenance and logistics.

b. *References.* Information in this manual is presented in general terms. For specific information, references are made to other PII manuals. See TM 9-1425-386-L, List of Applicable Publications (LOAP) for PERSHING II Field Artillery Missile System. See appendix A for publications that are referenced in this manual but not listed in TM 9-1425-386-L.

1-2. MAINTENANCE FORMS, RECORDS, AND REPORTS.

Department of the Army forms and procedures used for equipment maintenance will be those prescribed by DA PAM 738-750, The Army Maintenance Management System (TAMMS). The DA PAM is published in the Maintenance Management UPDATE. Units may subscribe to *Maintenance Management UPDATE* by submitting a completed DA Form 12-13.

1-3. NOMENCLATURE CROSS-REFERENCE LIST.

Shortened nomenclature is used in this manual to make the information easier for you to read. Table 1-1 contains a cross-reference between shortened nomenclature and official nomenclature.

Table 1-1. Nomenclature Cross-Reference List

Manual Nomenclature	Official Nomenclature
Battery control central or BCC	Battery Control Central, Guided Missile, Semitrailer Mounted; (PERSHING II)
CAS site power distribution equipment	CAS Site Power Distribution Equipment
Chemical agent alarm	Alarm Unit, Chemical Agent Automatic Alarm, M42
Chemical agent detector	Detector Unit, Chemical Agent Automatic Alarm, M43
Combat alert status van or CAS van	Combat Alert Status Facility, Guided Missile, Semitrailer Mounted
Contact team tool kit	Tool Kit, Guided Missile Maintenance: Contact Team (PERSHING II)
Digital repair tool kit	Tool Kit, Guided Missile Maintenance: Digital Repair (PERSHING II)
EL cover	Cover, Erector-Launcher
EL tractor	Truck Tractor: Tactical, 8x8, Heavy Expanded Mobility w/crane, M1001 or M983
Electrical repair shop	Electrical Repair Shop Equipment, Guided Missile System, Semitrailer Mounted
Electrical shop tool kit	Tool Kit, Guided Missile Maintenance: Electrical Shop (PERSHING II)

1-3. NOMENCLATURE CROSS-REFERENCE LIST – CONTINUED.

Table 1-1. Nomenclature Cross-Reference List – Continued

Manual Nomenclature	Official Nomenclature
Erector-launcher or EL	Erector-Launcher, Guided Missile, Semitrailer Mounted, M1003: (PERSHING II)
Firing site tool kit	Tool Kit, Guided Missile Assembler, Firing Site (PERSHING II)
First stage	Propulsion Section, Guided Missile, First Stage: (PERSHING II)
First stage aft skirt	Aft Skirt Assembly, First Stage
First stage hoisting beam	Beam, Hoisting, GM, First Stage: (PERSHING II)
Four-leg sling	Sling, 4-Leg, Missile Container and Shelter; (PERSHING II)
General support unit tool kit or GSU tool kit	Tool Kit, Guided Missile Maintenance: GSU, Class V (PERSHING II)
GIEU simulator	Simulator, Missile/Erector-Launcher/Ground Integrated Electronics Unit
Guidance and control/adapter or G&C/A section	Guidance and Control/Adapter Section, GM (PERSHING II)
Load test tool kit	Tool Kit, Guided Missile Maintenance: Test Equipment, Load Test, Ground Handling Equipment (PERSHING II)
Mechanical repair shop	Mechanical Repair Shop Equipment, Guided Missile System, Semitrailer Mounted
Mechanical shop tool kit	Tool Kit, Guided Missile Maintenance: Mechanical Shop (PERSHING II)
Mechanics and structures tool kit	Tool Kit, Guided Missile Maintenance: Mechanics and Structures (PERSHING II)
Missile section trainer (MASC)	Missile Section Trainer (MASC)
Missile section trainer (MT)	Missile Section Trainer (MT)
Missile simulator	Trainer Simulator, Electronic Module: (PERSHING II)
M1001, truck, tractor	Truck, Tractor: 10-Ton, 8x8, High Mobility, M1001
M1002, truck, wrecker	Truck, Wrecker: 10-Ton, 8x8, High Mobility, M1002
M200A1, trailer	Trailer, Generator Chassis, 1-1/2-Ton, 2-Wheel, M200A1
M35A2, truck	Truck, Cargo, 2-1/2-Ton, 6x6, M35A2
M818, tractor	Truck, Cargo, 5-Ton, 6x6, M818
M871, trailer	Semitrailer, Stake, 22-Ton, 4-Wheel, M871
M925, truck	Truck, Cargo, 5-Ton, Dropside, 6x6, M925
M928, truck	Truck, Cargo, 5-Ton, XLWB, W/W, 6x6, M928
M931, tractor	Truck, Tractor, 5-Ton, 6x6, M931
M932, tractor	Truck, Tractor, 5-Ton, 6x6, M932
M983, tractor	Truck, Tractor, Tactical, 10-Ton, 8x8, High Mobility, M983
M984, recovery vehicle	Truck, Recovery: Tactical, 10-Ton, 8x8, High Mobility, M984
Platoon control central or PCC	Platoon Control Central, Guided Missile, Transportable (PERSHING II)
Platoon control central trainer or PCC trainer	Trainer, Platoon Control Central, Guided Missile: (PERSHING II)
Radar section or RS	Radar Section, Guided Missile (PERSHING II)
Rear area power unit or RAPU	Generator Group, Guided Missile System, Trailer Mounted, OV-81/TSM-150 (XO-1)
Rear area SCTS filter box assembly	Filter Assembly, Commercial Power, SCTS
Reference scene generation facility or RSGF	Reference Scene Generation Facility: (PERSHING II)
Reference scene generation facility trainer or RSGF trainer	Trainer, Reference Scene Generation Facility, Dismounted (PERSHING II)
Repair parts shop	Repair Parts Shop, Guided Missile System, Semitrailer Mounted
S&A recode test set	Test Set, Safe and Arm Recoding
Second stage	Propulsion Section, Guided Missile, Second Stage (PERSHING II)
Second stage aft skirt	Aft Skirt Assembly, Second Stage
Second stage hoisting beam	Beam, Hoisting, GM, Second Stage: (PERSHING II)
Special tools and equipment tool kit	Tool Kit, Guided Missile Maintenance: Special Tools and Equipment (PERSHING II)
Storage maintenance tool kit	Tool Kit, Guided Missile Maintenance: Organizational Storage Maintenance (PERSHING II)

1-3. NOMENCLATURE CROSS-REFERENCE LIST – CONTINUED.

Table 1-1. Nomenclature Cross-Reference List – Continued

Manual Nomenclature	Official Nomenclature
Supply and packaging/preservation shop or P&P shop	Supply and Packaging/Preservation Shop, Guided Missile System, Semitrailer Mounted
System components test station or SCTS	Test Station, Guided Missile System, System Components, Semitrailer Mounted, AN/TSM-150 (XO-1)
System components test station trainer or SCTS trainer	Test Station, Guided Missile System, Training (PERSHING II)
Two-leg sling	Sling, 2-Leg, Container and Propulsion Section: (PERSHING II)
Universal sling	Sling, Reentry Vehicle: (PERSHING II)
Warhead section or WS	Warhead Section, Guided Missile: (PERSHING II)

1-4. REPORTING EQUIPMENT IMPROVEMENT RECOMMENDATIONS (EIR).

If your PERSHING II system needs improvement, let us know. Send us an EIR. You, the user, are the only one who can tell us what you don't like about your equipment. Let us know why you don't like the design. Put it on an SF 368 (Quality Deficiency Report). Mail it to Commander, U.S. Army Missile Command, ATTN: AMSMI-QA-QM-D, Redstone Arsenal, AL 35898-5290. We'll send you a reply.

1-5. LIST OF ABBREVIATIONS.

Table 1-2 contains an alphabetical list of abbreviations and symbols used in this manual, with their meanings. For a complete list of authorized abbreviations, see AR 310-50.

Table 1-2. List of Abbreviations

Abbreviation or Symbol	Meaning
A	ampere
ac	alternating current
AND	alphanumeric display
APOD	aerial port of debarkation
APOE	aerial port of embarkation
AR	Army Regulation
ASU	air servicer unit
ATS	automatic test set
BCC	battery control central
Btu	British thermal unit
CAS	combat alert status
CB	chemical/biological
CHEFU	clocked high energy firing unit
CONUS	Continental United States
CRT	cathode ray tube
cu ft	cubic foot
dB	decibel
dc	direct current
DCU	digital correlator unit
DS/GS	direct support/general support
EBW	exploding bridgewire

Table 1-2. List of Abbreviations – Continued

Abbreviation or Symbol	Meaning
ECCCS	European command and control console system
EL	erector-launcher
EME	electromagnetic effect
FSC	forward support company
ft	foot
G&C/A	guidance and control/adapter
GIEU	ground integrated electronics unit
GSE	ground support equipment
GSU	general support unit
HQ	headquarters
Hz	hertz
IEU	integrated electronics unit
ILA	interface logic assembly
IMS	inertial measurement system
in.	inch
kg	kilogram
kHz	kilohertz
km	kilometer
kPa	kilopascal

1-5. LIST OF ABBREVIATIONS – CONTINUED.

Table 1-2. List of Abbreviations – Continued

Abbreviation or Symbol	Meaning
kph	kilometer per hour
kW	kilowatt
l	liter
lb	pound
LCA	launch control assembly
LSC	linear shaped charge
m	meter
M&S	maintenance and supply
MASC	missile assembly simulated countdown
MHz	megahertz
mph	mile per hour
MT	modified tactical
NCS	nozzle control system
OIC	officer in charge
ORF	operational readiness float
P&P	packaging and preservation
PAC	PERSHING airborne computer
PAL	permissive action link
PCA	power control assembly
PCC	platoon control central
PE	protective entrance
PLC	PERSHING launch computer
POD	port of debarkation
POE	port of embarkation
psi	pounds per square inch

Table 1-2. List of Abbreviations – Continued

Abbreviation or Symbol	Meaning
psig	pounds-per-square-inch gage
PII	PERSHING II
qt	quart
RAPU	rear area power unit
RCS	reaction control system
RF	radio frequency
RGU	rate gyro unit
RLCU	remote launch control unit
RS	radar section
RSGF	reference scene generation facility
RV	reentry vehicle
S&A	safe and arm
SCTS	system components test station
SMDC	shielded mild detonating cord
TM	technical manual
TPD	transient protection device
UHF	ultrahigh frequency
UUT	unit under test
VCS	vane control system
WFS	warhead functional simulator
WS	warhead section
XLWB	extra long wheel base
°C	degree Celsius
°F	degree Fahrenheit
%	percent

1-6. SAFETY CONSIDERATIONS.

a. *General*. Personnel must be familiar with and observe all safety precautions when working on or handling the following:

- Electronic equipment
- High-pressure air systems
- Hazardous solvents
- Toxic or flammable chemicals.

During missile firing preparations, personnel in the area will be limited to those necessary to accomplish the task. Drilling equipment will not be used to modify items containing explosives. See AR 385-16 for other general safety precautions to be observed during training, service practice, and combat. See TM 9-1115-386-12&P for precautions on nuclear warhead operations and handling.

b. *Initial Blast Area* (fig. 1-1). To provide personnel safety from missile blast during firing, personnel are evacuated upon completion of their duties to a barricaded position at least 350 feet (106.7 m) from the missile. The launch key will not be turned until all personnel have reached the proper distance away from the missile. All loose and flammable items must be removed from the initial blast area before firing.

c. *Electrical Storms*. The missile on the erector-launcher (EL) is susceptible to lightning strikes. The possibility is greatly increased when the missile is vertical. During electrical storms, the missile should be lowered and personnel should be cleared from the immediate vicinity of the equipment. See TM 9-1300-206 for more information about operation in electrical storms.

1-6. SAFETY CONSIDERATIONS – CONTINUED.

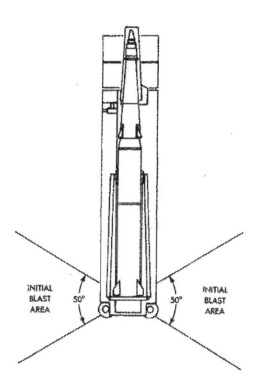

Figure 1-1. Initial Blast Area

d. *Misfire*. A misfire condition exists when a missile fails to lift off after the launch sequence INITIATE switch is turned. A misfire after missile battery activation is considered to be an unsafe condition. If a misfire occurs, the unit commander is notified immediately. The unit commander will direct the unit to perform the misfire procedures in TM 9-1425-386-10-2-1.

CHAPTER 2
EQUIPMENT DESCRIPTION

Section I. SYSTEM DESCRIPTION

2-1. EQUIPMENT CHARACTERISTICS, CAPABILITIES, AND FEATURES.

The PII missile system is a mobile, medium range, surface-to-surface missile system capable of quick reaction to various targets. The PII missile system's ability to quickly respond to various targets is due to many features. Some of these features are:

- Mobility
- Operating environment
- Technology
- Warhead options.

a. *Mobility.* The PII system is mounted on wheeled vehicles. It can operate on improved roads or off the road.

b. *Operating Environment.* The PII missile system will perform within a wide range of physical conditions, including:

- Limited visibility due to darkness, fog, or dust
- Humidity up to 90%
- Changes in ambient temperatures and atmospheric pressures
- Salt spray in coastal regions.

c. *Technology.* Computers in the missile and EL monitor and control all countdown functions.

(1) The PERSHING airborne computer (PAC) is the onboard missile computer. Before launch, it monitors critical missile functions and communicates with the PERSHING launch computer (PLC). After launch, the PAC controls in-flight operations such as targeting and stage separation.

(2) The PLC is mounted on the EL. Crewmembers monitor and control prelaunch functions using its keyboard and alphanumeric display (AND).

(3) Monitor and control equipment is protected from electromagnetic effects (EME).

d. *Warhead Options.* The airburst/surface burst (AB/SB) capability can be selected.

2-2. OPERATIONAL RANGE.

The two-stage PII missile is capable of delivering a payload to any range between 60 statute miles (97 km) and 1,080 statute miles (1,738 km).

2-3. MODES OF TRANSPORTATION.

The PII missile system is transportable by ground transport, air, rail, or ship. The following ground transport mode description is limited to forward area equipment.

a. *Ground Transport Mode.* The PII missile system uses various wheeled vehicles to transport the major end items. The forward area equipment is transported as follows:

- EL is towed by an M983 or M1001 tractor.
- Missile is transported assembled on the EL without the radar section (RS) and warhead section (WS).
- RS is carried in the EL pallet cradle.
- WS is carried either in its container on an M871 trailer towed by an M931 tractor or mated with the RS on the EL in the rotating pallet.
- Platoon control central (PCC) is mounted on an M928 cargo truck. The truck tows a trailer-mounted 30 kW generator. The generator supplies power to the PCC.
- Battery control central (BCC) is towed by an M931 tractor. BCC power is supplied by a trailer-mounted 30 kW generator towed by an M928 cargo truck.
- Combat alert status (CAS) van is towed by an M931 tractor.
- AN/TRC-184 radio terminal set is mounted on an M35A2 cargo truck that tows two 10 kW generators.
- Reference scene generation facility (RSGF) is mounted on an M928 cargo truck that tows a 30 kW generator. The generator supplies power to the RSGF.

See section III of this chapter for more information on ground transportation vehicles.

b. *Air Transport Mode.* The PII missile system can be transported by C-130, C-141, and C-5A aircraft.

c. *Rail Transport Mode.* The PII missile system can be transported on flat, gondola, or U.S. standard railroad cars.

d. *Ship Transport Mode.* The PII missile system can be transported on Mariner, Liberty, and Victory class ships.

2-4. TYPICAL OPERATION.

Operation is presented in general terms. Missile assembly and countdown are covered; firing role and mission are not covered.

a. *Missile Assembly.* The missile sections arrive in shipping containers (see paragraph 2-8 for container description). The missile can be assembled with containers on trailers as shown in figure 2-1 or 2-2. The missile also can be assembled with containers removed from the trailers and placed on the ground as shown in figure 2-3 or 2-4. The missile sections are removed from the containers and assembled on the EL. The WS may be transported either mated to the RS in the EL pallet or in the WS container.

For detailed missile assembly procedures, see TM 9-1425-386-10-3-1. For detailed missile disassembly procedures, see TM 9-1425-386-10-3-2. For information on emplacement within limited terrain, see TM 9-1425-386-10-2-1.

b. *Countdown.* The following events occur at a firing position.

(1) Emplacement:
- Vehicles parked
- Cables connected
- Power source started
- EL leveled
- EL tractor moved.

2-4. TYPICAL OPERATION -- CONTINUED.

(2) WS mating:

- WS placed on EL pallet (if required)
- WS electrically mated
- WS mechanically mated.

(3) Prelaunch operations:

- EL control
- Inertial measurement system (IMS) warmup
- Targeting
- EL checkout
- IMS ready
- Missile power on
- Accumulator charged
- Digital correlator unit (DCU) test
- IMS alinement
- WS presets (optional)
- Confidence complete
- Hot hold (optional)
- Intent words entered
- Ignition enable entered
- Permissive action link (PAL) code entered
- Appropriate EL cover panels removed.

(4) Launch operations:

- Remote selected
- Launch area evacuated
- Launch sequence initiated
- Missile erected
- Launch sequence
- Lift-off.

(5) Evacuation.

For more information on countdown operations, see TM 9-1425-399-24-1.

2-4. TYPICAL OPERATION – CONTINUED.

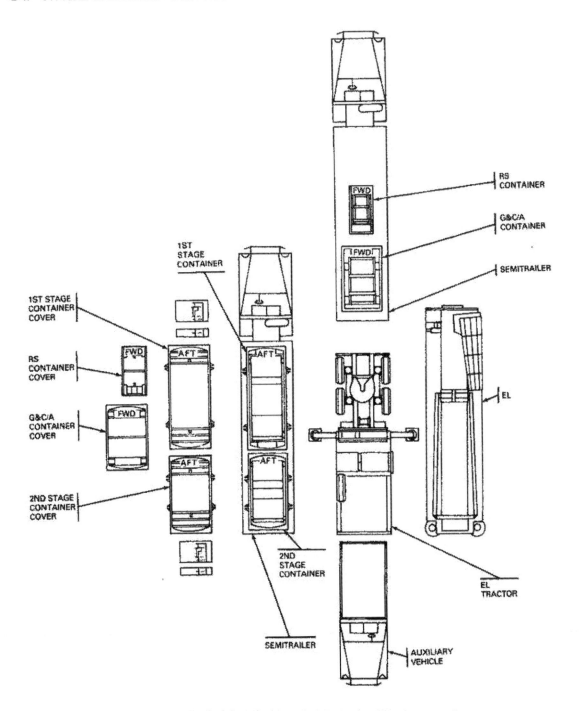

Figure 2-1. Typical Positioning of Vehicles for Missile Assembly
(EL Tractor to Roadside of EL)

2-4. TYPICAL OPERATION – CONTINUED.

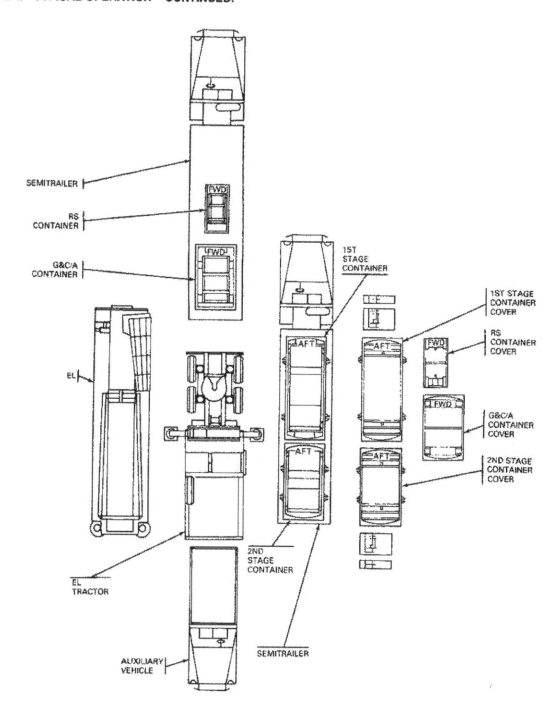

**Figure 2-2. Typical Positioning of Vehicles for Missile Assembly
(EL Tractor to Curbside of EL)**

2-4. TYPICAL OPERATION – CONTINUED.

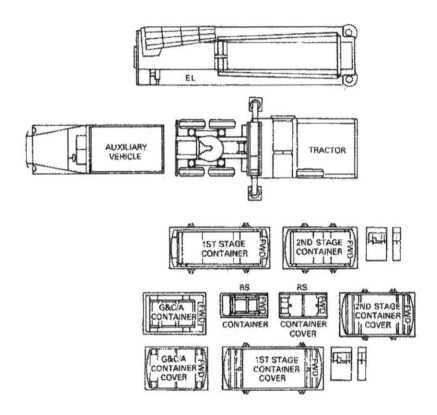

Figure 2-3. Typical Positioning of Containers for Missile Assembly
(EL Tractor to Roadside of EL)

2-4. TYPICAL OPERATION – CONTINUED.

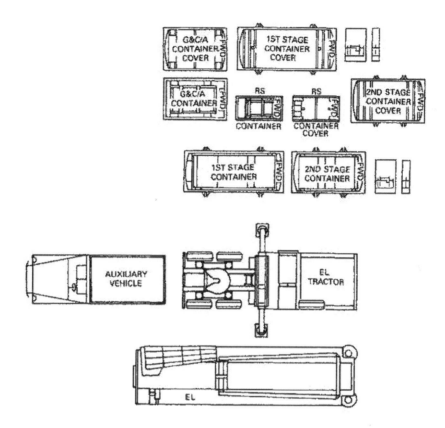

Figure 2-4. Typical Positioning of Containers for Missile Assembly
(EL Tractor to Curbside of EL)

2-5. TYPICAL TRAJECTORY.

A typical PII missile trajectory is shown in figure 2-5. The three trajectory phases are:

a. Boost Phase. This phase occurs from lift-off until the end of powered flight. During this time the PAC controls:

- First stage separation after burnout
- Second stage ignition
- Second stage separation and thrust termination
- Steering and guidance.

b. Midcourse Phase. This phase occurs from the end of powered flight (thrust termination) until reentry to the atmosphere. During this time, the PAC controls the vane control system (VCS) and reaction control system (RCS). These aid guidance of the reentry vehicle (RV).

c. Terminal Phase. This phase starts at atmospheric reentry and continues until warhead detonation. The PAC compares stored target data with data obtained from radar. Using this target data comparison, the PAC controls the RV fins for final guidance.

2-5. TYPICAL TRAJECTORY – CONTINUED.

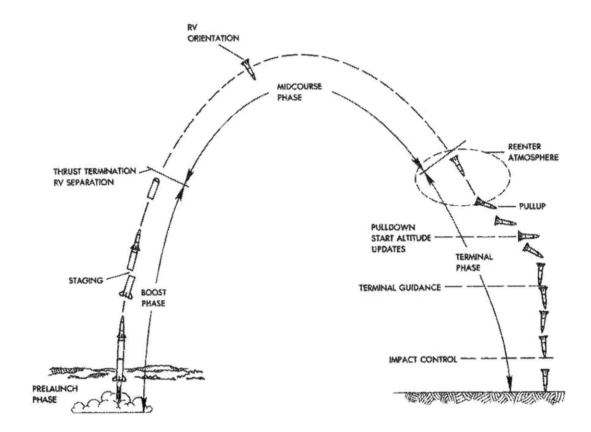

Figure 2-5. Typical PII Missile Trajectory

Section II. MISSILE DESCRIPTION

2-6. MISSILE.

The PII missile is a two-stage, five-section aerodynamic structure. It is 417.55 inches (10.6 m) long and 40 inches (1.0 m) in diameter, and weighs approximately 16,451 pounds (7,462 kg). Each missile section is interchangeable with any other like section in the PII missile system. Missile section weights and dimensions are given in chapter 5. Figure 2-6 shows the PII missile with a view of each section.

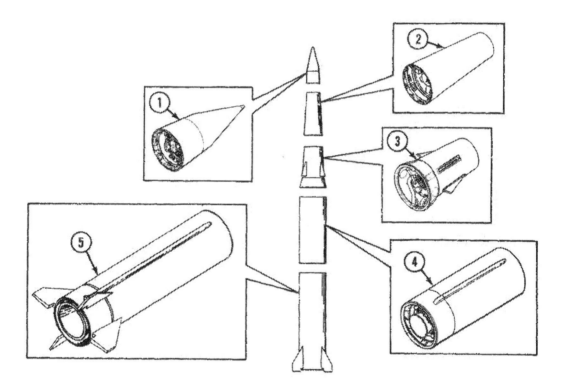

Figure 2-6. PII Missile

1 RS. Provides target and altitude information.
2 WS. Contains warhead.
3 G&C/A. Provides guidance and control functions.
4 SECOND STAGE. Contains second stage, solid-propellant motor.
5 FIRST STAGE. Contains first stage, solid-propellant motor.

2-7. MISSILE SECTIONS.

a. *First Stage.* The two main assemblies of the first stage are the aft skirt assembly and rocket motor assembly. Figure 2-7 shows first stage major features.

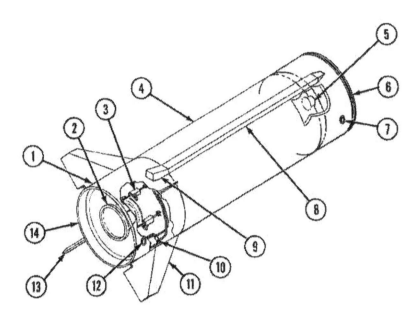

Figure 2-7. First Stage Major Features

1 AFT SKIRT ASSEMBLY. Cylindrical aluminum assembly that contains aft lift points, VCS, nozzle control system (NCS), aft attach ring, and cables.

2 MOVABLE NOZZLE. Directs thrust developed by rocket motor during first stage operation. Nozzle provides pitch and yaw control during first stage powered flight.

3 NCS. Controls movement of nozzle and provides nozzle position data to PAC.

4 ROCKET MOTOR ASSEMBLY. Filament-wound cylindrical assembly that contains forward lift points, solid propellant, and first stage ignition system. Rocket motor assembly also serves as outer surface of forward section of first stage.

5 FIRST STAGE IGNITION SYSTEM. Allows electrical ignition of first stage rocket motor and prevents inadvertent launch. Ignition system contains igniter, safe and arm (S&A) device, initiators, clocked high energy firing unit (CHEFU), and high voltage cables.

6 FORWARD ATTACH RING. Allows mating of first stage to second stage.

7 FORWARD LIFT POINT. Two lift points allow attachment of first stage hoisting beam so that first stage can be lifted and moved.

8 CONDUIT COVER ASSEMBLY. Externally mounted cover that routes cables from aft skirt assembly, lengthwise along outside of rocket motor assembly, to inside forward skirt.

9 TAILPLUG CONNECTORS. Allow electrical interface between missile and ground integrated electronics unit (GIEU) on EL.

10 VCS. Controls movement of two movable fins and provides fin position data to PAC.

11 MOVABLE FIN. Two movable fins located opposite each other on the first stage aft skirt. Fins provide roll control during first stage powered flight.

2-7. MISSILE SECTIONS – CONTINUED.

12 AFT LIFT POINT. Two lift points allow attachment of first stage hoisting beam so that first stage can be lifted and moved.

13 FIXED FIN. Two fixed fins located opposite each other on the first stage aft skirt. Fins provide stability during first stage powered flight.

14 AFT ATTACH RING. Allows mating of first stage to azimuth ring assembly on EL.

   b. Second Stage. The two main assemblies of the second stage are the aft skirt assembly and rocket motor assembly. Figure 2-8 shows second stage major features.

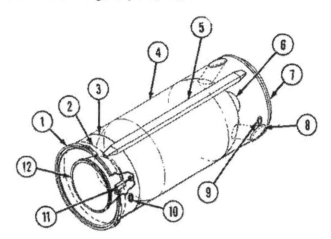

Figure 2-8. Second Stage Major Features

1 AFT ATTACH RING. Allows mating of first stage to second stage.

2 FIRST STAGE SEPARATION SYSTEM. Allows separation of first stage from second stage after first stage burnout and before second stage ignition. Separation system contains linear shaped charge (LSC) separation ring, detonators, CHEFU, and high voltage cables.

3 AFT SKIRT ASSEMBLY. Cylindrical aluminum assembly that contains aft lift points, NCS, aft splice ring, and cables.

4 ROCKET MOTOR ASSEMBLY. Filament-wound cylindrical assembly that contains forward lift points, solid propellant, and second stage ignition system. Rocket motor assembly also serves as outer surface of forward section of second stage.

5 CONDUIT COVER ASSEMBLY. Externally mounted cover that routes cables from aft skirt assembly, lengthwise along outside of rocket motor assembly, to inside forward skirt.

6 SECOND STAGE IGNITION SYSTEM. Allows electrical ignition of second stage rocket motor. Ignition system contains igniter, initiators, CHEFU, and high voltage cables.

7 FORWARD ATTACH RING. Allows mating of second stage to G&C/A.

8 THRUST REVERSAL SYSTEM. Allows second stage reverse thrust to be developed after RV separation so that second stage will not interfere with RV flight. Thrust reversal system contains three thrust reversal ports, LSC rings, shielded mild detonating cord (SMDC), thrust reversal manifold, detonators, CHEFU, and high voltage cables.

9 FORWARD LIFT POINT. Two lift points allow attachment of second stage hoisting beam so that second stage can be lifted and moved.

10 AFT LIFT POINT. Two lift points allow attachment of second stage hoisting beam so that second stage can be lifted and moved.

11 NCS. Controls movement of nozzle and provides nozzle position data to PAC.

12 MOVABLE NOZZLE. Directs thrust developed by rocket motor during second stage operation. Nozzle provides pitch and yaw control during second stage powered flight.

2-7. MISSILE SECTIONS – CONTINUED.

c. G&C/A. The G&C/A consists of the guidance and control (G&C) and the adapter assemblies. Figure 2-9 shows G&C/A major features.

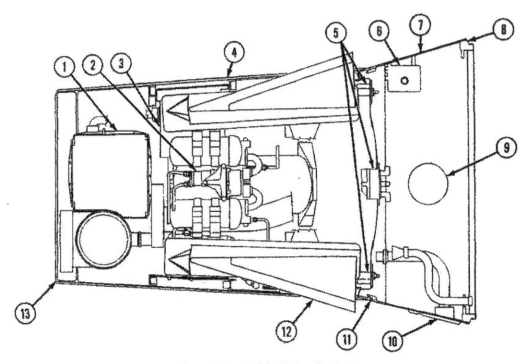

Figure 2-9. G&C/A Major Features

1   INTEGRATED ELECTRONICS UNIT (IEU). Controls all functions of missile during flight. IEU contains PAC, DCU and IMS.

2   VCS. Controls movement of four fins and provides fin position data to PAC. VCS operates during second stage powered flight and terminal portion of flight.

3   MISSILE BATTERY. Provides electrical power to electrical/electronic assemblies during flight.

4   G&C SUPPORT STRUCTURE. Conical aluminum assembly wrapped with ablative heatshield. Support structure provides mount and protection for internal components.

5   RCS. Provides pitch, yaw, and roll control during midcourse portion of flight.

6   TWO-AXIS RATE GYRO UNIT (RGU). Provides pitch and yaw data to PAC during boost portion of flight.

7   ADAPTER SUPPORT STRUCTURE. Conical aluminum assembly with heat-resistant coating. Support structure provides mount and protection for internal components.

8   AFT ATTACH RING. Allows mating of second stage to G&C/A.

9   ORDNANCE ACCESS COVER. Allows access to second stage motor initiators, separation detonators, and thrust reversal detonators.

10   UMBILICAL COVER. Allows automatic closing of RV ground cooling system ducts during flight.

11   RV SEPARATION SYSTEM. Allows separation of RV from adapter/second stage at end of powered flight. Separation system contains LSC separation ring, detonators, CHEFU, and high voltage cables.

12   G&C/A FIN. Four fins provide roll control during second stage powered flight and pitch, yaw, and roll control during terminal portion of flight.

13   QUICK ACCESS SPLICE RING. Allows mating of G&C/A to WS.

2-7. MISSILE SECTIONS – CONTINUED.

   d. WS. Figure 2-10 shows WS major features.

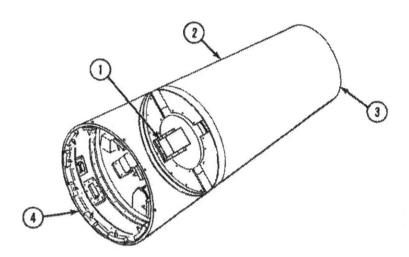

Figure 2-10. WS Major Features

   1   THREE-AXIS RGU. Provides roll control information during boost phase of flight; provides pitch, yaw, and roll control information during midcourse and terminal phases of flight.
   2   WS SUPPORT STRUCTURE. Conical aluminum alloy assembly covered with ablative material.
   3   QUICK ACCESS SPLICE RING. Allows mating RS to WS.
   4   QUICK ACCESS SPLICE SEGMENT. Ten segments allow mating WS to G&C/A.

For more information on the WS, see TM 9-1115-386-12&P.

2-7. MISSILE SECTIONS – CONTINUED.

e. *RS.* The two main assemblies of the RS are the support structure and the radome. Figure 2-11 shows RS major features.

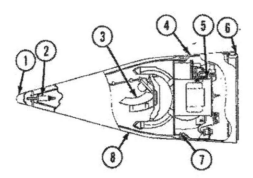

Figure 2-11. RS Major Features

1 NOSE CAP. Seals off forward end of radome and provides protection during reentry.

2 IMPACT FUZE. Used to detonate warhead in surface burst option.

3 STABILIZED ANTENNA. Allows radar unit to transmit and receive radio frequency (RF) energy.

4 SUPPORT STRUCTURE. Conical aluminum assembly wrapped with an ablative heatshield.

5 RADAR UNIT. Provides target site information to PAC for comparison with stored target site information.

6 QUICK ACCESS SPLICE SEGMENT. Eight splice segments allow mating the RS to the WS.

7 IMPACT FUZE. Four fuzes used to detonate warhead in surface burst option.

8 RADOME. Reinforced glass/epoxy shell that covers radar unit antenna. It also acts as heatshield.

2-8. SHIPPING AND STORAGE CONTAINERS.

Each PII missile section has a container used for shipping and storage. The containers are similar, differing mostly in size and weight. Figure 2-12 shows PII shipping and storage containers. Each container has three main areas:

- Base assembly that houses and supports the section
- Suspension system that protects the section from shock
- Cover that completes the enclosure.

Each container also has the same general features:

- Can be transported by truck, aircraft, rail, or ship
- Has covers that are removable with use of slings and hoist
- Has storage space for equipment records (except first or second stage aft skirt container)
- Can be handled with forklift or slings
- Can be towed for short distances on skids
- Can withstand varied environmental conditions
- Has provisions for testing missile section in container (except first or second stage aft skirt container)
- Has EME protection (except first or second stage aft skirt container).

For more information on shipping and storage containers, see TM 9-8140-395-14.

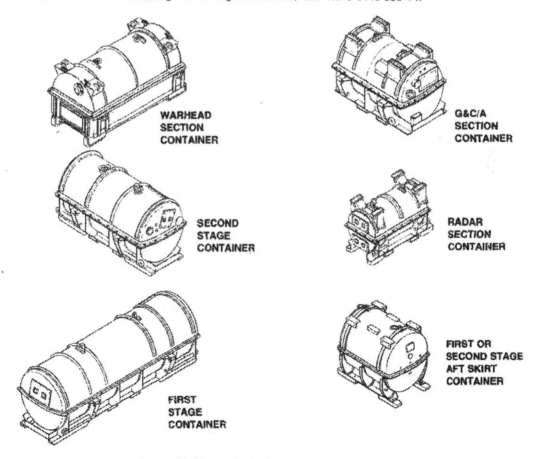

Figure 2-12. PII Shipping and Storage Containers

2-9. MISSILE SECTION PROTECTIVE COVERS AND SHIELDS.

Eight protective covers are used on the missile sections. They provide protection from either the environment or EME. Some also provide safety to personnel and equipment. All covers can be installed or removed quickly. Figure 2-13 shows the protective covers and shields.

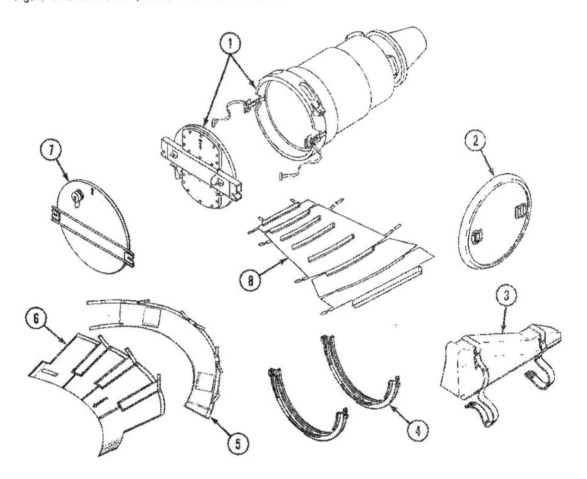

Figure 2-13. Missile Section Protective Covers and Shields

1 RS PROTECTIVE COVER SET. Stored on RS. Cover set consists of:

   • RS protective cover – provides RF protection to personnel during RS test portion of countdown; protects radome and allows RS to be lifted.

   • RS shield assembly – installed on aft end of RS except when RS is mated to WS; provides EME protection.

2 G&C/A SHIELD ASSEMBLY. Installed on forward end of G&C/A when G&C/A is on EL for travel; provides EME protection. Shield assembly is stored on EL.

3 FIN PROTECTIVE COVER. Four covers protect G&C/A fins when G&C/A is on EL or in container; provide protection from physical damage.

4 LSC PROTECTIVE COVER. Covers LSC separation ring of G&C/A during testing in system components test station (SCTS); provides personnel protection from inadvertent firing of RV separation system. Cover is stored in SCTS.

2-9. MISSILE SECTION PROTECTIVE COVERS AND SHIELDS -- CONTINUED.

5 ADAPTER PROTECTIVE COVER. Covers adapter portion of G&C/A when G&C/A is on EL; protects from scratches, nicks, etc.

6 G&C/A PROTECTIVE COVER. Covers G&C portion of G&C/A when G&C/A is on EL; protects ablative surface from scratches, nicks, etc.

7 WS SHIELD ASSEMBLY. Covers rear of WS when WS is mated to RS; provides EME protection. Shield assembly is stored on EL.

8 SUNSHIELD. Installed over G&C/A, WS, and RS in sunny weather to prevent heat absorption.

Section III. GROUND SUPPORT EQUIPMENT

2-10. ERECTOR-LAUNCHER (EL).

The EL is a semitrailer used as a platform to assemble, transport, count, erect, and launch the PERSHING II missile. It is capable of operation either in a platoon or alone. The EL is equipped with a tandem bogie and high flotation tires so that it can operate over rough terrain and muddy or sandy soil. For a detailed description of the EL, see TM 9-1440-389-10. Figure 2-14 shows EL major features.

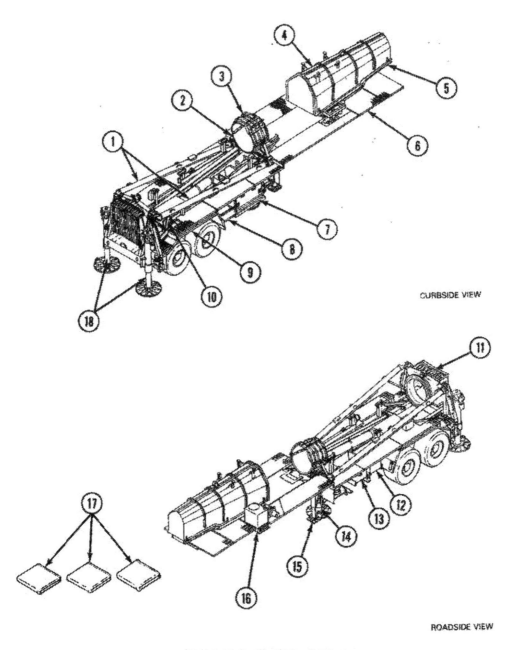

CURBSIDE VIEW

ROADSIDE VIEW

Figure 2-14. EL Major Features

2-10. ERECTOR-LAUNCHER (EL) – CONTINUED.

1 BOOMS. Support missile cradle during erection and recapture of missile.
2 MISSILE CRADLE. Supports missile during transport, erection, and recapture of missile.
3 RETAINING RING SEGMENTS. Used to retain missile in missile cradle during transport.
4 EL PALLET COVER. Protects RS and WS during travel.
5 EL PALLET. Platform used to transport and mate WS and RS.
6 WORK PLATFORM. Work area for mating RV sections.
7 GIEU. Consists of LCA and PCA with protective door.
8 HYDRAULIC CONTROL PANEL. Contains controls and indicators for system hydraulic function.
9 UPLOCK RELEASE MECHANISM. Releases azimuth ring uplock allowing missile recapture.
10 UPLOCK ASSEMBLY. Locks azimuth ring in erect (firing) position.
11 AZIMUTH RING ASSEMBLY. Consists of launch platform, blast deflector, and ring for mating missile.
12 EL POWER SUPPLY. Provides 28 V dc power to EL.
13 MISSILE POWER SUPPLY. Provides 28 V dc power to missile.
14 FRONT JACK. Used to raise, lower, and level front of EL.
15 LANDING GEAR. Supports front of EL when EL is detached from tractor and not supported by jacks.
16 HYDRAULIC OIL TANK. Nonpressurized reservoir for hydraulic oil.
17 PROTECTIVE COVERS. Protect G&C/A and RS.
18 REAR JACKS. Used to raise, lower, and level rear of EL.

2-10.1. EL COVER.

The EL cover includes 22 side panels and 16 top panels, which are secured to frames mounted to the EL. A single panel is mounted on the blast deflector. Individual panels can be removed quickly for EL maintenance or for missile erection. The EL cover provides environmental protection for the first stage and second stage missile sections and part of the G&C/A section. For a detailed description of the EL cover, see TM 9-1440-389-10. Figure 2-14.1 shows the EL cover.

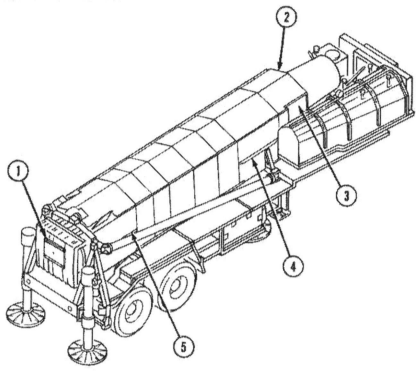

Figure 2-14.1. EL Cover

1   AFT PANEL. Remains in place except for maintenance.
2   TOP PANELS. Panels with half hinges removed for missile erection, and panels with rollers stored behind side panels for missile erection.
3   FORWARD SIDE PANELS. Curbside and roadside panels folded outward for missile erection.
4   SIDE PANELS. Curbside and roadside panels removed for missile erection.
5   AFT SIDE PANELS. Curbside and roadside panels removed for missile erection.

2-11. EL TRACTOR.

The EL tractor is a 10-ton, 8x8, diesel-powered vehicle. It can pull the EL with the missile over improved surfaces, such as roads or highways, or over rough terrain. Power steering assists tractor operator in steering four front wheels. Either the M1001 tractor or the M983 tractor can be used as an EL tractor.

a. *M1001 Tractor*. Figure 2-15 shows M1001 tractor major features. For more information on the M1001 tractor, see TM 9-2320-282-10.

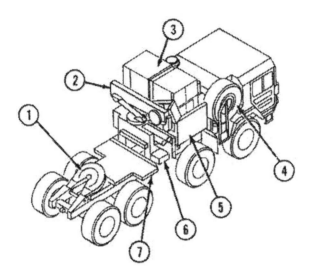

Figure 2-15. M1001 Tractor Major Features

1   FIFTH WHEEL. Allows connecting EL to tractor.
2   CRANE. Used to hoist missile sections for assembly and disassembly.
3   30 KW GENERATOR. Supplies power to EL during tactical operation.
4   SPARE TIRE. For EL tractor.
5   PLATFORM. Folds down to provide access to 30 kW generator controls.
6   POWER DISTRIBUTION BOX. Has connectors for connecting cable from 30 kW generator to EL.
7   PLATFORM. Used to remove RS cover.

2-11. EL TRACTOR – CONTINUED.

b. *M983 Tractor.* Figure 2-16 shows M983 tractor major features. For more information on the M983 tractor, see TM 9-2320-279-10.

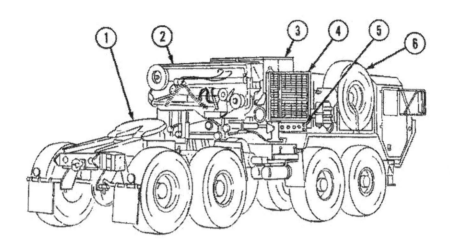

Figure 2-16. M983 Tractor Major Features

1   FIFTH WHEEL. Allows connecting EL to tractor.
2   CRANE. Used to hoist missile sections for assembly and disassembly.
3   30 KW GENERATOR. Supplies power to EL during tactical operation.
4   PLATFORM. Folds down to provide access to 30 kW generator controls.
5   POWER DISTRIBUTION BOX. Has connectors for connecting cable from 30 kW generator to EL.
6   SPARE TIRE. For EL tractor.

2-12. TRANSPORTATION VEHICLES.

The vehicles used to transport the PII missile sections and ground support equipment can travel over improved roads or rough terrain. Transportation vehicles include:

- M871 trailer
- M818 5-ton tractor
- M931 5-ton tractor
- M932 5-ton tractor
- M35A2 2 1/2-ton cargo truck
- M928 5-ton cargo truck
- M925 5-ton dropside cargo truck
- M1002 10-ton recovery vehicle
- M984 10-ton recovery vehicle.

a. *M818, M931, and M932 Tractors/M871 Trailer.* The M818, M931, and M932 tractors are 5-ton, 6x6, diesel-powered tow trucks. They tow the 22-ton M871 trailer, BCC, CAS van, SCTS, electrical repair shop, mechanical repair shop, supply and packaging/preservation shop, and repair parts shop. The M871 trailer carries PII missile sections in their containers. For more information on the M931 and M932 tractors, see TM 9-2320-272-10. For more information on the M871 trailer, see TM 9-2330-358-14&P. For more information on the M818 tractor, see TM 9-2320-260-10-1. Figure 2-17 shows the M818, M931, and M932 tractors/M871 trailer.

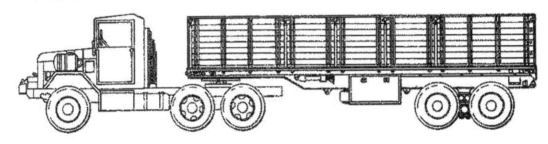

Figure 2-17. M818, M931, and M932 Tractors/M871 Trailer

b. *M35A2 Cargo Truck.* The M35A2 is a 2-1/2-ton, 6x6, diesel-powered cargo truck used as a general supply vehicle for the firing platoon. For more information on the M35A2 truck, see TM 9-2320-209-10-1. Figure 2-18 shows the M35A2 cargo truck.

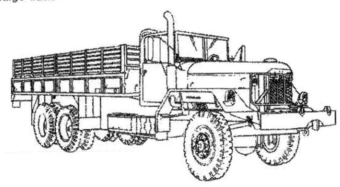

Figure 2-18. M35A2 Cargo Truck

2-12. TRANSPORTATION VEHICLES – CONTINUED.

    c. M928 Cargo Truck. The M928 is a 5-ton, 6x6, extra-long wheelbase (XLWB) diesel-powered cargo truck. The PCC and RSGF are mounted on M928 cargo trucks. For more information on the M928 truck, see TM 9-2320-272-10. Figure 2-19 shows the M928 cargo truck.

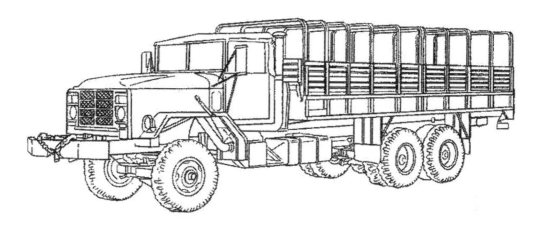

Figure 2-19. M928 Cargo Truck

    d. M925 Dropside Cargo Truck. The M925 is a 5-ton, dropside, 6x6, diesel-powered cargo truck used to tow the RAPU and to transport forward support company (FSC) equipment. For more information on the M925 truck, see TM 9-2320-272-10. Figure 2-20 shows the M925 dropside cargo truck.

Figure 2-20. M925 Dropside Cargo Truck

2-12. TRANSPORTATION VEHICLES – CONTINUED.

e. M1002 Recovery Vehicle. The M1002 is a 10-ton, 8x8, diesel-powered truck similar to the M1001 tracto except for:

- No fifth wheel
- No 30 kW generator
- Large storage box for supplies and tools to service PII system vehicles
- Recovery winch and crane for towing disabled PII system vehicles.

In some areas, the M984 recovery vehicle is used in place of the M1002. For more information on the M100 recovery vehicle, see TM 9-2320-282-10. Figure 2-21 shows the M1002 recovery vehicle.

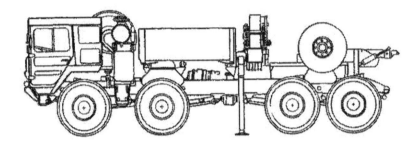

Figure 2-21. M1002 Recovery Vehicle

f. M984 Recovery Vehicle. The M984 is a 10-ton, 8x8, diesel-powered truck similar to the M983 tractor excep for:

- No fifth wheel
- No 30 kW generator
- Large storage box for supplies and tools to service PII system vehicles
- Recovery winch and crane for towing disabled PII system vehicles.

In some areas, the M1002 recovery vehicle is used in place of the M984. For more information on the M98 recovery vehicle, see TM 9-2320-279-10. Figure 2-22 shows the M984 recovery vehicle.

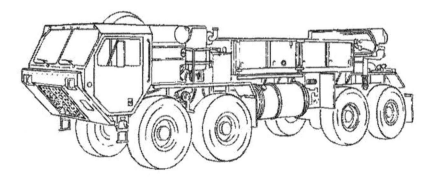

Figure 2-22. M984 Recovery Vehicle

2-13. PLATOON CONTROL CENTRAL (PCC).

The PCC is a mobile, self-contained command center for a PII platoon. The PCC provides a secure location to control and monitor the countdown and firing of up to three PII missiles. The PCC equipment is housed in an S-280 facilitized electrical equipment shelter mounted on an M928 cargo truck. Power is supplied to the PCC by a trailer-mounted 30 kW generator. For more information on the PCC, see TM 9-1430-392-10.

   a. *Exterior Features.* Figure 2-23 shows PCC major exterior features.

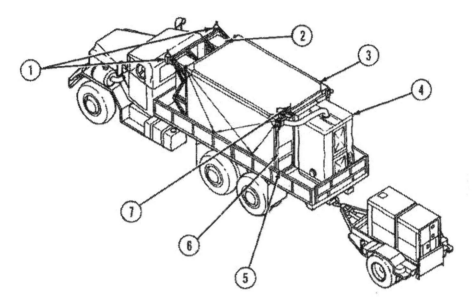

Figure 2-23. PCC Major Exterior Features

1   ANTENNA MAST BASES. Mount radio communication antennas.
2   AIR CONDITIONERS. Two units, each capable of 18,000 Btu cooling, 12,000 Btu heating, and ventilation.
3   FACILITIZED SHELTER. Contains PCC equipment.
4   PROTECTIVE ENTRANCE (PE). Provides 5-minute air bath to remove CB contaminants from personnel before entering PCC.
5   POWER ENTRY PANEL. Connects power, ground, and telephone cables to PCC.
6   SIGNAL ENTRY PANEL. Connects signal cables to PCC.
7   ELECTRICAL HORN. Alerts personnel that launch is about to take place.

2-13.   PLATOON CONTROL CENTRAL (PCC) – CONTINUED.

b. *Interior Features.* Figure 2-24 shows PCC major interior features.

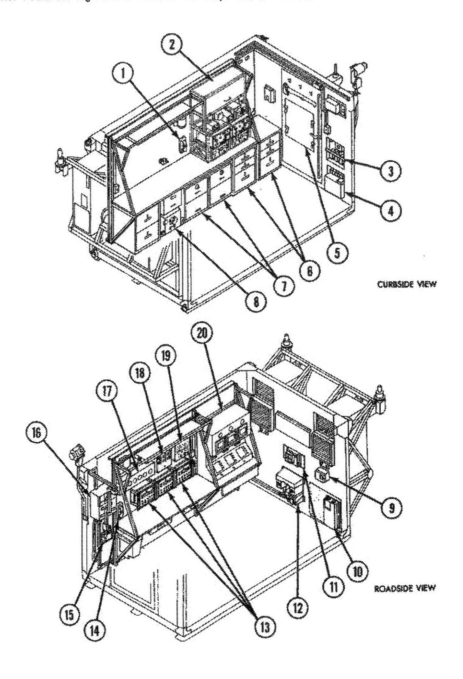

Figure 2-24. PCC Major Interior Features

2-13. PLATOON CONTROL CENTRAL (PCC) – CONTINUED.

1   TELEPHONE TA-312/PT. Used for telephone communications.

2   COMMUNICATIONS EQUIPMENT RACK. Used to mount radio communications equipment.

3   SIGNAL ENTRY PANEL. Used to connect internal signal cables to external signal cables.

4   POWER ENTRY PANEL. Used to connect power distribution panel to exterior power source, internal ground to external grounding cables, telephones to external lines.

5   EMERGENCY PANEL. Provides emergency exit.

6   STORAGE DRAWERS. Store items needed for operation and maintenance, such as countdown tapes and cables.

7   SECURITY CABINETS. Drawers have removable storage cases that can be secured when removed.

8   REMOVABLE SAFE. Stores material and can be removed to secure area.

9   CHEMICAL AGENT ALARM. Alerts personnel when detector triggers alarm unit.

10   CONVERTER. Changes input voltage to 208 V 400 Hz for operation of CB equipment.

11   CONTROL MODULE. Provides single point from which CB system status is monitored.

12   FORWARD ENTRY PANEL. Connects external filter unit, air conditioners, and antennas to internal components.

13   REMOTE LAUNCH CONTROL UNITS (RLCU's). Provide launch capability for three missiles.

14   TELEPHONE TA-312/PT. Used for telephone communications.

15   CHEMICAL AGENT DETECTOR. Triggers alarm unit when chemical agents are detected.

16   POWER DISTRIBUTION PANEL. Provides circuit breaker-protected ac and dc power to components.

17   REMOTE CONTROL CONSOLE. When installed, used with communications system.

18   STATUS DISPLAY PANEL. Displays status of missiles.

19   LAUNCH WINDOW PANEL. Calculates and displays launch window times for three missiles.

20   ILA. Connects each missile with PCC; controls printers, AND's, and status display panel; and provides communications between EL's and PCC.

2-14. BATTERY CONTROL CENTRAL (BCC).

The BCC is a mobile, self-contained command center for a PII battery. The BCC provides a secure location to control and monitor the countdown and firing of up to three PII missiles. The BCC also provides a secure operating location for command and control of the battery. The BCC equipment is housed in a modified M1006 van towed by an M931 tractor. Power is supplied to the BCC by a trailer-mounted 30 kW generator. For more information on the BCC, see TM 9-1430-397-14.

2-14. BATTERY CONTROL CENTRAL (BCC) – CONTINUED.

a. *Exterior Features*. Figure 2-25 shows BCC major exterior features.

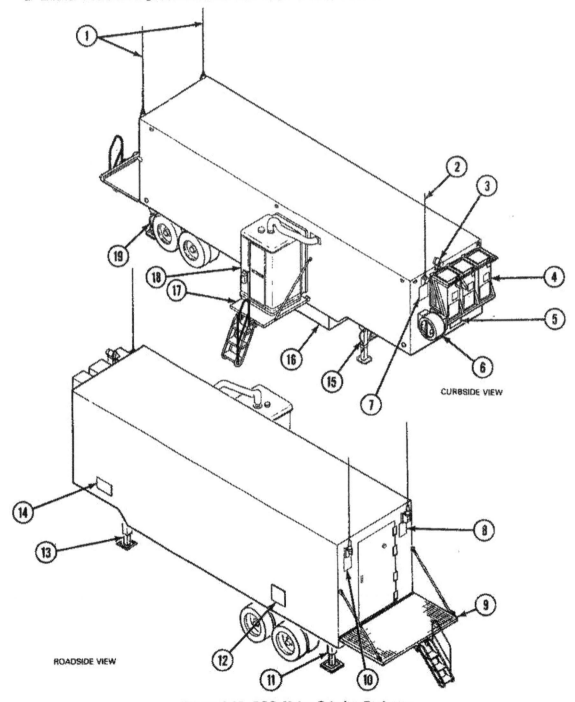

Figure 2-25. BCC Major Exterior Features

2-14. BATTERY CONTROL CENTRAL (BCC) – CONTINUED.

1 ANTENNAS AS-1729/VRC. Whip antennas for radio communications.
2 ANTENNA. Whip antenna for radio communications.
3 ELECTRICAL HORN. Used to alert personnel that launch is about to take place.
4 AIR CONDITIONER. Three units, each capable of 18,000 Btu cooling, 12,000 Btu heating, and ventilation.
5 MOUNTING PANEL A2. Used to connect air conditioners and emergency dc power.
6 FILTER UNIT. Contains gas and particulate filters with blower to force filtered air into BCC.
7 MOUNTING PANEL A6. Used to connect whip antenna, filter unit, and electrical horn.
8 MOUNTING PANEL A11. Used to connect antenna cables.
9 REAR WORK PLATFORM. Provides work area.
10 MOUNTING PANEL A12. Used to connect antenna cables.
11 LEVELING JACK. Levels BCC during emplacement.
12 SIGNAL ENTRY PANEL A15. Used to connect BCC to EL's of platoon 1, radio terminal set AN/TRC-184, and remote antennas.
13 LANDING GEAR. Supports front of BCC.
14 POWER ENTRY PANEL A1. Used to connect power, ground, satellite communications terminal AN/MSC-64, CB alarm, switchboard, and facsimile machine cables.
15 LANDING GEAR. Supports front of BCC.
16 STORAGE CABINET. Storage location for PE and other loose items.
17 SIDE PERSONNEL PLATFORM. Platform for PE.
18 PE. Provides 5-minute air bath to remove CB contaminants from personnel before entering BCC.
19 LEVELING JACK. Levels BCC during emplacement.

2-14. BATTERY CONTROL CENTRAL (BCC) – CONTINUED.

b. *Interior Features.* Figure 2-26 shows BCC major interior features.

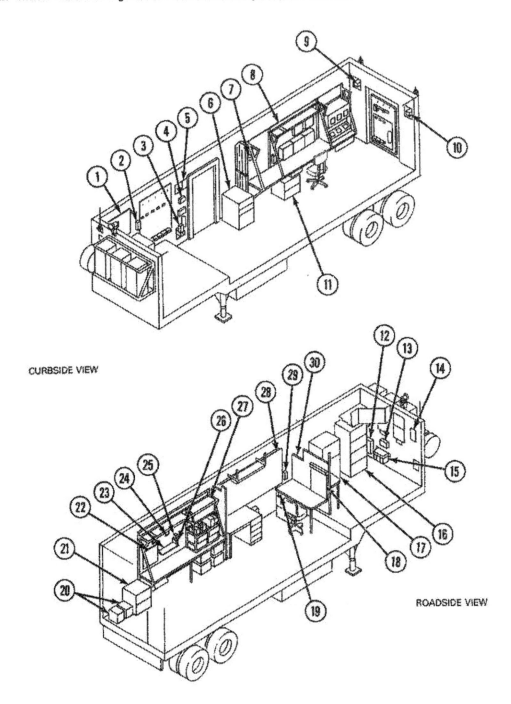

CURBSIDE VIEW

ROADSIDE VIEW

Figure 2-26. BCC Major Interior Features

2-14. BATTERY CONTROL CENTRAL (BCC) – CONTINUED.

1 PLOTTER BOARD. Used to record tactical situation data.

2 TELEPHONE TA-312/PT. Used for telephone communications.

3 CHEMICAL AGENT DETECTOR M43. Triggers alarm unit when chemical agents are detected.

4 CHEMICAL AGENT ALARM M42. Alerts personnel when detector triggers alarm unit.

5 PE BLAST AIR VALVE. Controls air flow to PE.

6 STORAGE CABINET. Stores items needed for operation and maintenance, such as countdown tapes and cables.

7 RECEPTACLE BOX. Used to connect satellite communications terminal AN/MSC-64 equipment when installed.

8 COMMAND AND LAUNCH CONTROL CONSOLE. Provides command and launch control capability.

9 SIGNAL ENTRY PANEL. Used to connect internal signal cables to external cables.

10 SIGNAL ENTRY PANEL. Used to connect internal signal cables to external cables.

11 SECURITY CABINET. Used to store material.

12 CONVERTER. Changes input voltage to 208 V 400 Hz for operation of CB equipment.

13 CONTROL MODULE. Provides single point from which CB system status is monitored.

14 MOUNTING PANEL A6. Used to connect external filter unit, whip antenna, and horn to internal components.

15 MOUNTING PANEL A2. Used to connect external filter unit, air conditioners, and backup dc power to internal components.

16 FILE CABINET. Used for document storage.

17 POWER DISTRIBUTION CABINET. Contains panel that provides circuit breaker-protected ac and dc power to components.

18 STATUS BOARD. Used to display battery status information.

19 SECURITY CABINET. Used to store material.

20 ELECTRIC SPACE HEATERS. Two portable heaters used for auxiliary heating of BCC.

21 SECURITY CABINET. Drawers have removable storage cases that can be secured when removed.

22 ERASABLE BOARD. Used for crew notes.

23 BOOKSHELF. Used for book storage.

24 REMOTE CONTROL CONSOLE. When installed, used with communications system.

25 TELEPHONE BRACKET. Provides capability for additional telephone TA-312/PT.

26 TELEPHONE TA-312/PT. Used for telephone communications.

27 COMMUNICATIONS EQUIPMENT RACK. Used to mount radio communications equipment.

28 MAPBOARD. Used to display battery maps.

29 TELEPHONE TA-312/PT. Used for telephone communications.

30 SWITCHBOARD RECEPTACLE BOX. Used to connect switchboard when installed.

2-15. COMBAT ALERT STATUS (CAS) VAN.

The CAS van is a mobile, self-contained command center. It provides a secure location to control and monitor the countdown and firing of up to nine PII missiles. The CAS equipment is housed in a modified M1006 van towed by an M931 tractor. Power is applied to the CAS van by CAS site power distribution equipment. For more information on the CAS van, see TM 9-1430-398-14.

2-15. COMBAT ALERT STATUS (CAS) VAN – CONTINUED.

a. *Exterior Features.* Figure 2-27 shows CAS van major exterior features.

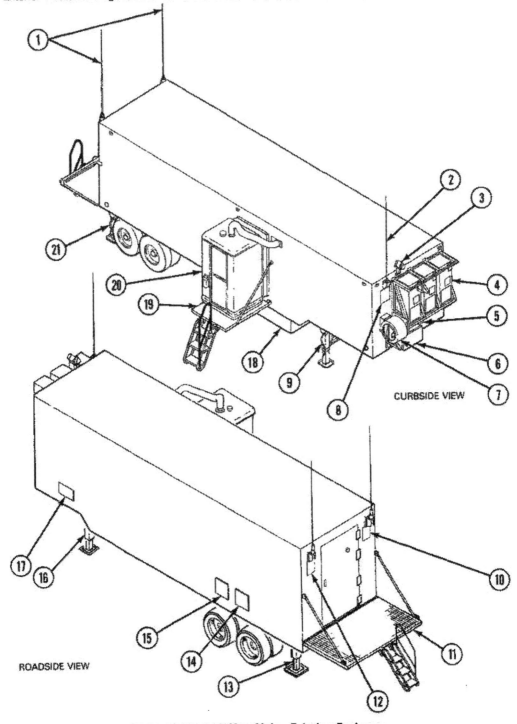

CURBSIDE VIEW

ROADSIDE VIEW

Figure 2-27. CAS Van Major Exterior Features

2-15. COMBAT ALERT STATUS (CAS) VAN – CONTINUED.

1   ANTENNAS AS-1729/VRC. Whip antennas for radio communications.

2   ANTENNA. Whip antenna for radio communications.

3   ELECTRICAL HORN. Used to alert personnel that launch is about to take place.

4   AIR CONDITIONER. Three units, each capable of 18,000 Btu cooling, 12,000 Btu heating, and ventilation.

5   MOUNTING PANEL A2. Used to connect air conditioners and backup batteries.

6   BATTERY BOX. Houses two 12 V batteries for backup power.

7   FILTER UNIT. Contains gas and particulate filters with blower to force filtered air into CAS van.

8   MOUNTING PANEL A6. Used to connect whip antenna, filter unit, and electrical horn.

9   LANDING GEAR. Supports front of van.

10   MOUNTING PANEL A11. Used to connect antenna cables.

11   REAR WORK PLATFORM. Provides work area.

12   MOUNTING PANEL A12. Used to connect antenna cables.

13   LEVELING JACK. Levels van during emplacement.

14   SIGNAL ENTRY PANEL A14. Used to connect CAS van to EL's of platoon 1, radio terminal set AN/TRC-184, and remote antennas.

15   SIGNAL ENTRY PANEL A15. Used to connect CAS van to EL's of platoons 2 and 3.

16   LANDING GEAR. Supports front of van.

17   POWER ENTRY PANEL A1. Used to connect power, ground, European Command and Control Console System (ECCCS), satellite communications terminal AN/MSC-64, switchboard, and facsimile machine cables.

18   STORAGE CABINET. Storage location for PE and other loose items.

19   SIDE PERSONNEL PLATFORM. Platform for PE.

20   PE. Provides 5-minute air bath to remove CB contaminants from personnel before entering van.

21   LEVELING JACK. Levels van during emplacement.

2-15. COMBAT ALERT STATUS (CAS) VAN – CONTINUED.

b. *Interior Features.* Figure 2-28 shows CAS van major interior features.

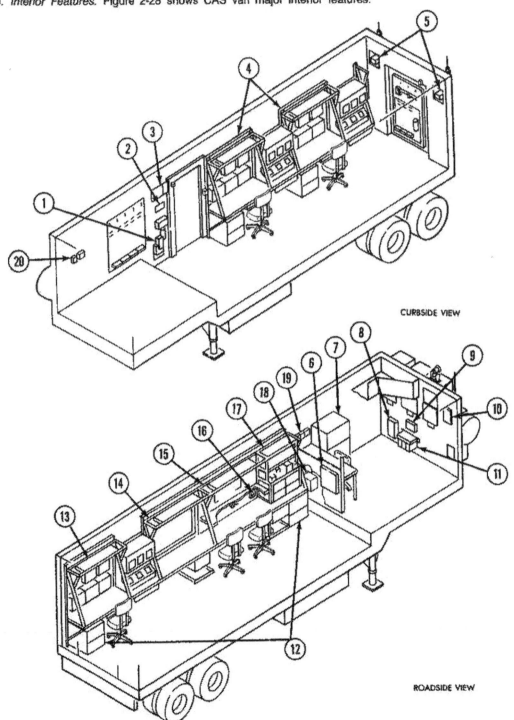

CURBSIDE VIEW

ROADSIDE VIEW

Figure 2-28. CAS Van Major Interior Features

2-15. COMBAT ALERT STATUS (CAS) VAN -- CONTINUED.

1 CHEMICAL AGENT DETECTOR M43. Triggers alarm unit when chemical agents are detected.

2 CHEMICAL AGENT ALARM M42. Alerts personnel when detector triggers alarm unit.

3 PE BLAST AIR VALVE. Controls air flow to PE.

4 COMMAND AND LAUNCH CONTROL CONSOLES. Each console provides command and launch control capability for three missiles.

5 SIGNAL ENTRY PANEL. Used to connect internal signal cables to external cables.

6 STATUS BOARD. Used to display battery status information.

7 POWER DISTRIBUTION CABINET. Contains panel that provides circuit breaker-protected ac and dc power to components.

8 CONVERTER. Changes input voltage to 208 V 400 Hz for operation of CB equipment.

9 CONTROL MODULE. Provides single point from which CB system status is monitored.

10 MOUNTING PANEL A6. Used to connect filter unit, whip antenna, and horn to internal components.

11 MOUNTING PANEL A2. Used to connect filter unit, air conditioners, and backup batteries to internal components.

12 SECURITY CABINETS. Used to store material.

13 COMMAND AND LAUNCH CONTROL CONSOLE. Provides command and launch control capability for three missiles.

14 MAPBOARD. Used to display battery maps.

15 REMOTE CONTROL CONSOLE. When installed, used with communication system.

16 TELEPHONE TA-312/PT. Used for telephone communications.

17 COMMUNICATIONS EQUIPMENT RACK. Used to mount radio communications equipment.

18 ELECTRIC SPACE HEATER. Two portable heaters used for auxiliary heating of van.

19 SWITCHBOARD RECEPTACLE BOX. Used to connect switchboard when installed.

20 RECEPTACLE BOX. Used to connect ECCCS equipment when installed.

2-16. S&A RECODE TEST SET.

The S&A recode test set allows testing or changing codes of the first stage ignition enable system. The test set is battery-powered and portable. It can be used with the first stage on the EL or in its container. For more information on the S&A recode test set, see TM 9-4935-387-14.

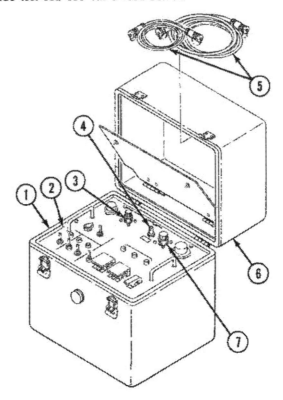

Figure 2-29. S&A Recode Test Set Major Features

1   CASE. Contains power supply, printed wiring board, and circuits necessary to code or recode first stage ignition S&A system.
2   CONTROL PANEL. Contains controls for self-test, code, and recode procedures.
3   BATTERY ADAPTER. Shorting adapter that connects to BATTERY CONNECTOR for test set operation.
4   GROUNDING COUPLER. Provides grounding during code and recode of S&A device when grounding cable is attached to test set.
5   SPECIAL PURPOSE CABLE ASSEMBLIES. Output cable assembly that connects to first stage ignition enable device; battery cable assembly that connects to battery charger.
6   COVER. Case cover and storage location for special purpose cable assemblies.
7   SELF-TEST ADAPTER. Self-test adapter that connects to OUTPUT CONNECTOR during self-test.

2-17. GROUND HANDLING EQUIPMENT.

Ground handling equipment is used for handling and mating missile sections and aft skirts and handling batteries. Ground handling equipment is used with the crane on the M1001 tractor for field handling, with an overhead crane for maintenance handling, and with a forklift for handling batteries. First and second stage hoisting beams and two-leg, four-leg, and universal slings are carried on the 5-ton auxiliary vehicle. All other handling equipment is located at maintenance facilities. Figure 2-30 shows ground handling equipment. For more information on ground handling equipment, see TM 9-1450-396-14.

2-17. GROUND HANDLING EQUIPMENT – CONTINUED.

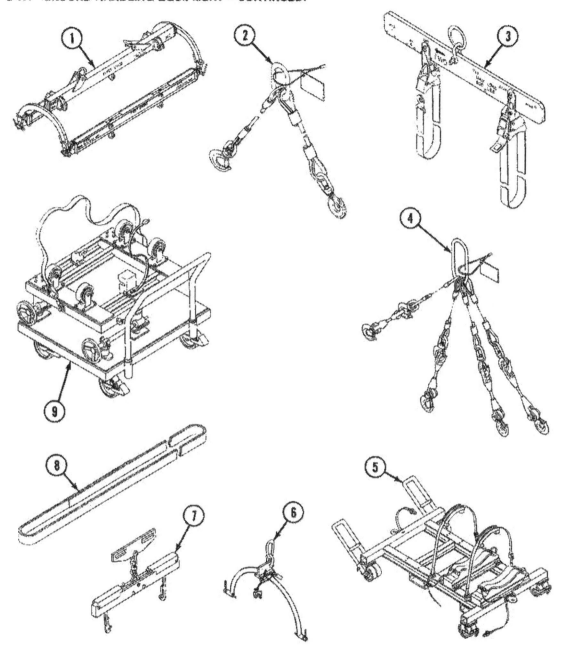

Figure 2-30. Ground Handling Equipment

1 FIRST STAGE HOISTING BEAM. Used with two-leg sling to lift first stage. First stage and second stage hoisting beams are alike except for physical size and lifting capacity.

2 TWO-LEG SLING. Used to lift all missile section container covers; used to lift first stage and second stage hoisting beams.

2-17. GROUND HANDLING EQUIPMENT – CONTINUED.

3   UNIVERSAL SLING. Adapts to allow lifting of G&C/A, WS, or RS.

4   FOUR-LEG SLING ASSEMBLY. With leg extensions, used to lift PCC, RSGF, and all missile section containers.

5   GUIDED MISSILE SECTION CRADLE. Used to support first or second stage missile section; used with aft skirt holding fixture to mate aft skirt with first or second stage missile section.

6   AFT SKIRT TURNOVER SLING. Used to lift aft skirt from wooden container and place aft skirt on aft skirt holding fixture.

7   BATTERY LIFTING SLING. Used with forklift to handle batteries mounted on tractors and CAS van.

8   AFT SKIRT CHOKER SLING. Used in handling of first or second stage aft skirt.

9   AFT SKIRT HOLDING FIXTURE. Used to support first or second stage aft skirt.

2-18. RADIO TERMINAL SET AN/TRC-184.

Radio terminal set AN/TRC-184, an upgraded version of the AN/TRC-133A, is a high frequency, single sideband radio set. It is made up of AN/GRC-193 receiver-transmitters and necessary control equipment. The equipment is housed in a modified S-280 shelter mounted on a 2-1/2-ton truck. The AN/TRC-184 is powered by two 10 kW generators. These are mounted on a trailer towed by the 2-1/2-ton truck. Figure 2-31 shows radio terminal set AN/TRC-184. For more information on the AN/TRC-184, see TM 11-5280-940-14.

Figure 2-31. Radio Terminal Set AN/TRC-184

2-19. SATELLITE COMMUNICATIONS TERMINAL AN/MSC-64.

Satellite communications terminal AN/MSC-64 is a tactical UHF satellite communications system. It can transmit and receive hard copy information. The equipment is housed in an S-615 shelter mounted on a 2-1/2-ton truck. Figure 2-32 shows satellite communications terminal AN/MSC-64. For more information on the AN/MSC-64, see TM 11-5895-1104-10.

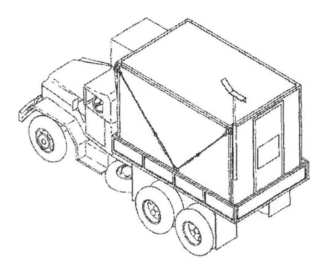

Figure 2-32. Satellite Communications Terminal AN/MSC-64

2-20. ELECTRICAL REPAIR SHOP.

The electrical repair shop is a self-contained mobile shop used for operator, organizational, and DS/GS testing and maintenance of electrical components and assemblies used to support PII operations. The shop is housed in an M373A2 van. Power is provided by a 60 kW generator mounted on a RAPU. For more information on the electrical repair shop, see TM 9-4935-394-14.

2-20. ELECTRICAL REPAIR SHOP – CONTINUED.

a. *Exterior Features*. Figure 2-33 shows electrical repair shop major exterior features.

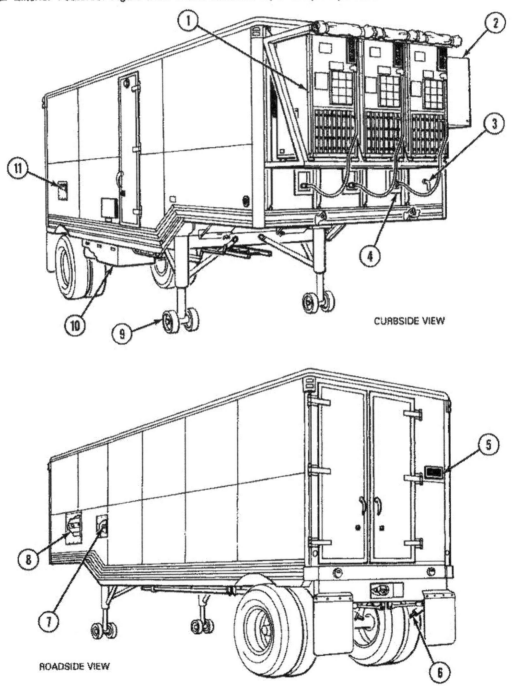

CURBSIDE VIEW

ROADSIDE VIEW

Figure 2-33. Electrical Repair Shop Major Exterior Features

2-20. ELECTRICAL REPAIR SHOP – CONTINUED.

1 AIR CONDITIONER. Three units, each capable of 18,000 Btu cooling, 12,000 Btu heating, and ventilation.
2 AIR COMPRESSOR. Provides up to 125 psig air for test and pneumatic tools.
3 AIR RECEIVER BLEED LINE. Exhausts compressed air and moisture from air receiver.
4 CONNECTOR PANEL. Three connector panels supply 3-phase 208 V ac 50/60 Hz power to air conditioners.
5 AIR VENT. When opened from inside, provides filtered, outside air.
6 LEVELING JACK AND FOOTPAD. Two leveling jacks and footpads support and level rear end of shop.
7 AUXILIARY CABLE ENTRY PANEL. Provides connection for 50 Hz commercial power source to shop.
8 MAIN CABLE ENTRY PANEL. Used to connect power and ground cables and field telephone lines.
9 LANDING GEAR. Supports front end of shop when disconnected from towing tractor.
10 STORAGE BOX. Storage for power and ground cables and for ground rods.
11 AIR SUPPLY PANEL. Provides external air supply connections.

2-20. ELECTRICAL REPAIR SHOP – CONTINUED.

b. *Interior Features*. Figure 2-34 shows electrical repair shop major interior features.

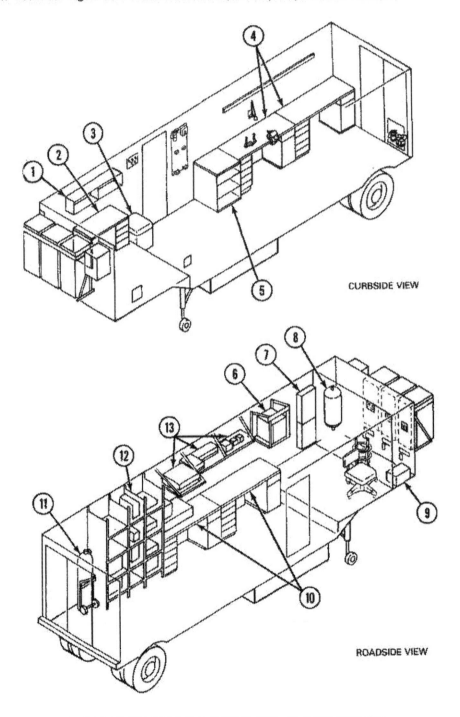

Figure 2-34. Electrical Repair Shop Major Interior Features

2-20. ELECTRICAL REPAIR SHOP – CONTINUED.

1 BOOKSHELF. Storage for technical manuals and other documents.
2 DESK. Used for general office purposes.
3 FILTER/TPD TEST SET AND STORAGE CASE. Used to test electrical filters and transient protection devices.
4 WORKBENCHES. Provide work area for electrical repair. One bench has bench vise and chassis vise. Both provide storage for parts and tools in drawers and cabinets under benches.
5 STORAGE SHELF. Three-shelf general purpose storage area.
6 28 V DC POWER SUPPLY. Provides power for test equipment.
7 ELECTRICAL DISTRIBUTION BOX. Provides control and distribution of circuit breaker-protected ac power to lights, convenience outlets, oven, and air conditioners in shop. Also contains connections for field telephones.
8 AIR RECEIVER. 20-gallon tank used with air compressor to provide stable compressed air for pneumatic tools.
9 TOOL BOX. Used for storing and transporting tools.
10 WORKBENCHES. Provide work area for electrical repair. Also provides storage for electronic counter and voltmeter on top of bench and storage for tools and parts in drawers and cabinets under benches.
11 PORTABLE GAS SERVICING UNIT. Used to service missile section nitrogen bottles and EL accumulator.
12 STORAGE RACK. Provides storage for tool boxes, tool kits, IMS multical tape cartridge, and test equipment including automatic cable tester, cable shield resistance test set, EBW grid dip meter, and status display panel test set.
13 TEST EQUIPMENT. Electronic test equipment including decade resistance box, igniter circuit tester, multimeters, and oscilloscope.

2-21. MECHANICAL REPAIR SHOP.

The mechanical repair shop is a self-contained mobile shop used for operator, organizational, and DS/GS testing and maintenance of mechanical components and assemblies used to support PII operations. The shop is housed in an M373A2 van. Power is provided by a 60 kW generator mounted on a RAPU. For more information on the mechanical repair shop, see TM 9-4935-394-14.

2-21. MECHANICAL REPAIR SHOP – CONTINUED.

a. *Exterior Features.* Figure 2-35 shows mechanical repair shop major exterior features.

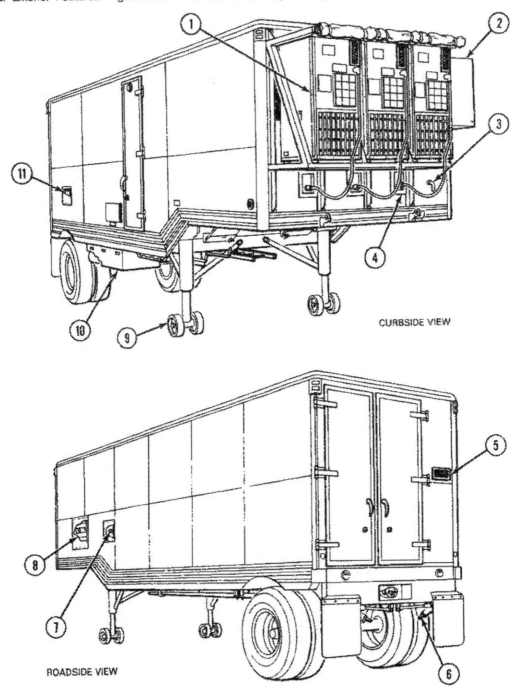

CURBSIDE VIEW

ROADSIDE VIEW

Figure 2-35. Mechanical Repair Shop Major Exterior Features

2-21. MECHANICAL REPAIR SHOP -- CONTINUED.

1. AIR CONDITIONER. Three units, each capable of 18,000 Btu cooling, 12,000 Btu heating, and ventilation.
2. AIR COMPRESSOR. Provides up to 125 psig air for test and pneumatic tools.
3. AIR RECEIVER BLEED LINE. Exhausts compressed air and moisture from air receiver.
4. CONNECTOR PANEL. Three connector panels supply 3-phase 208 V ac 50/60 Hz power to air conditioners.
5. AIR VENT. When opened from inside, provides filtered, outside air.
6. LEVELING JACK AND FOOTPAD. Two leveling jacks and footpads support and level rear end of shop.
7. AUXILIARY CABLE ENTRY PANEL. Provides connection for 50 Hz commercial power source to shop.
8. MAIN CABLE ENTRY PANEL. Used to connect power and ground cables and field telephone lines.
9. LANDING GEAR. Supports front end of shop when disconnected from towing tractor.
10. STORAGE BOX. Storage for power and ground cables and for ground rods.
11. AIR SUPPLY PANEL. Provides external air supply connections.

2-21. MECHANICAL REPAIR SHOP – CONTINUED.

b. *Interior Features.* Figure 2-36 shows mechanical repair shop major interior features.

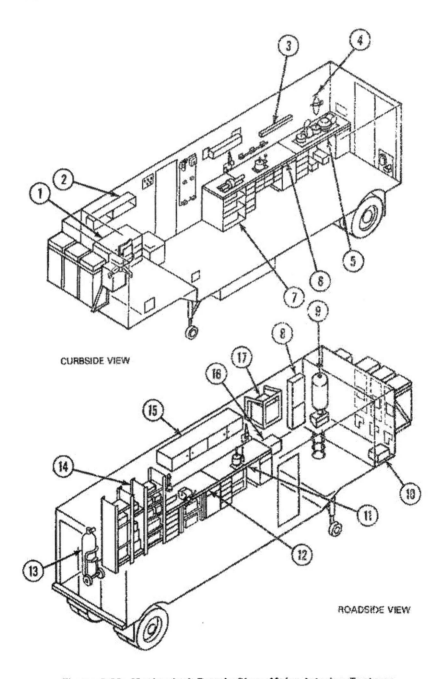

CURBSIDE VIEW

ROADSIDE VIEW

Figure 2-36. Mechanical Repair Shop Major Interior Features

2-21. MECHANICAL REPAIR SHOP – CONTINUED.

1. DESK. Used for general office purposes.

2. BOOKSHELF. Storage for technical manuals and other documents.

3. STORAGE BOX. Used to store turbine exhaust tube alinement tool.

4. REGULATOR/FILTER. Used to control air from receiver.

5. WORKBENCH. Used for hydrostatic testing of pneumatic hose assemblies and hose repair. Also provides storage for tools and parts in drawers and cabinet under bench.

6. WORKBENCH. Used for grinding and pressfitting parts and general repair. Also provides storage for tools and parts in drawers and cabinets under bench.

7. STORAGE RACK. Used to store leveling jack continuity test set.

8. ELECTRICAL DISTRIBUTION BOX. Provides control and distribution of circuit breaker-protected ac power to lights, convenience outlets, oven, and air conditioners in shop. Also contains connections for field telephones.

9. AIR RECEIVER. 20-gallon tank used with air compressor to provide stable compressed air for pneumatic tools and hydrostatic tester.

10. TOOL BOX. General purpose tool box.

11. WORKBENCH. Provides work area for drilling and mechanical repair. Also provides storage for tools and parts in drawers and cabinets under bench.

12. WORKBENCH. Provides work area with vise for mechanical repair. Also provides storage for tools and parts in drawers and cabinets under bench.

13. PORTABLE GAS SERVICING UNIT. Used to service missile section nitrogen bottles and EL accumulator.

14. STORAGE RACK. Provides storage for conduit winching tool, pressure gage tester, RS and WS carriage compression tool, screw-thread insert (heli-coil) repair kit, torque wrench analyzer tool set, and tool boxes.

15. OVERHEAD STORAGE CABINET. Used for storing tools and parts.

16. HYDRAULIC JACK TOOL SET. Used to bend and straighten damaged structural components.

17. 28 V DC POWER SUPPLY. Provides 28 V dc for leveling jack continuity test set operation.

2-22. SUPPLY AND PACKAGING/PRESERVATION (P&P) SHOP.

The P&P shop is a self-contained supply office and shop used for packaging and preservation of components required to support PII operations. The shop is housed in an M373A2 semitrailer van. Required electrical power is provided by a 30 kW generator mounted on a trailer. Shop temperature and humidity are controlled by two 18,000 Btu air conditioners. Connections are provided for field telephones. For more information on the P&P shop, see TM 9-4935-395-14.

a. *Exterior Features.* Figure 2-37 shows P&P shop major exterior features.

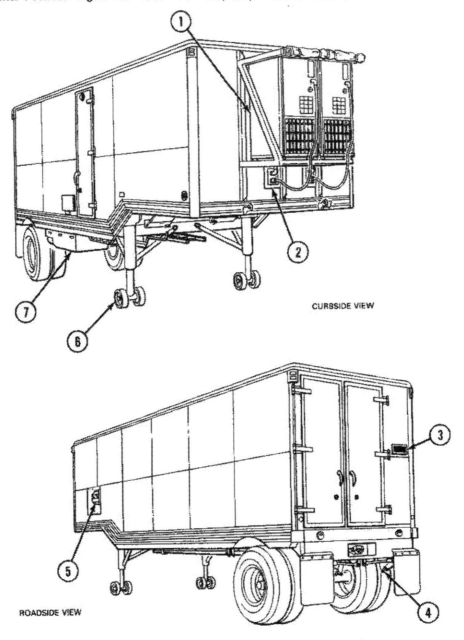

CURBSIDE VIEW

ROADSIDE VIEW

Figure 2-37. P&P Shop Major Exterior Features

2-22. SUPPLY AND PACKAGING/PRESERVATION (P&P) SHOP – CONTINUED.

1   AIR CONDITIONER. Two units, each capable of 18,000 Btu cooling, 12,000 Btu heating, and ventilation.
2   CONNECTOR PANEL. Two connector panels supply 3-phase 208 V ac 50/60 Hz power to air conditioners.
3   AIR VENT. When opened from inside, provides filtered outside air.
4   LEVELING JACK AND FOOTPAD. Two leveling jacks and footpads support and level rear end of shop.
5   CABLE ENTRY PANEL. Used to connect power and ground cables and field telephone lines.
6   LANDING GEAR. Supports forward end of shop when disconnected from towing tractor.
7   STORAGE BOX. Storage for power and ground cables and for ground rods.

2-22. SUPPLY AND PACKAGING/PRESERVATION (P&P) SHOP – CONTINUED.

b. *Interior Features.* Figure 2-38 shows P&P shop major interior features.

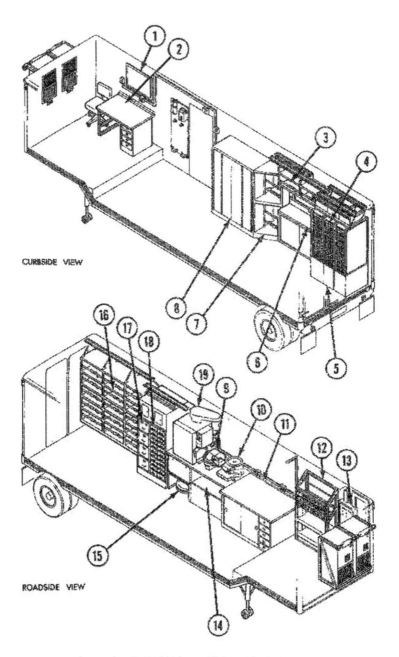

Figure 2-38. P&P Shop Major Interior Features

2-22. SUPPLY AND PACKAGING/PRESERVATION (P&P) SHOP -- CONTINUED.

1 BULLETIN BOARD. Office-type board used to post general information.

2 DESK. Used for general office procedures.

3 BOOKRACK. Contains shelves and webbing straps for storage and transportation of contents.

4 VISUAL CARD INDEX FILE. Contains 81 pullout slide card index files used to store supply records.

5 VACUUM PUMP STORAGE AREA. Single-door storage area for portable vacuum pump used to remove air during packaging procedures.

6 WORKBENCH. General purpose workbench with storage area.

7 BARREL RACK. Contains shelves and webbing straps for storage and transportation of two barrels of desiccant.

8 STORAGE CABINET. Two-shelf, general purpose storage cabinet with webbing straps.

9 TAPE DISPENSER. Electrically operated dispenser of hot tape in lengths from 6 to 36 inches, selected in 3-inch increments.

10 STENCIL CUTTING MACHINE. Manually-operated, 1/2-inch, single-character (number or letter) stencil cutter.

11 WORKBENCH. General purpose workbench with storage area for wrecking bar, saw, steel strapping tensioners, and steel strapping sealer.

12 PAPER RACK. Contains two each paper roll holders, paper cutters, and spare paper rolls.

13 ELECTRICAL DISTRIBUTION BOX. Provides control and distribution of circuit breaker-protected ac power to lights, convenience outlets, oven, and air conditioners in shop. Also contains connections for field telephones.

14 STENCIL WASTE DRAWER. Catches and holds punch-out waste from stencil cutting machine.

15 STEEL STRAPPING MACHINE. Circular dispenser mounted on floor next to card file drawers. Feeds out flat steel strapping.

16 BOOKRACK. Three bookracks contain shelves with webbing straps for storage and transportation of contents.

17 FILE SAFE. Four-drawer combination safe for storage of materials.

18 CARD FILE. Card file with 2 drawers for supply record storage.

19 OVEN. Single-phase 115 V 50/60 Hz thermostatically controlled oven used for drying, preheating, aging, and curing. Heats up to 300° F.

2-23. REPAIR PARTS SHOP.

The repair parts shop is a self-contained mobile facility used for storage of spare assemblies, subassemblies, and spare parts required to support PII operations. The shop is housed in an M373A2 semitrailer van. Required electrical power is provided by a 30 kW generator mounted on a trailer. Shop temperature and humidity are controlled by two 18,000 Btu air conditioners. Connections are provided for field telephones. For more information on the repair parts shop, see TM 9-4935-395-14.

2-23. REPAIR PARTS SHOP – CONTINUED.

a. *Exterior Features*. Figure 2-39 shows repair parts shop major exterior features.

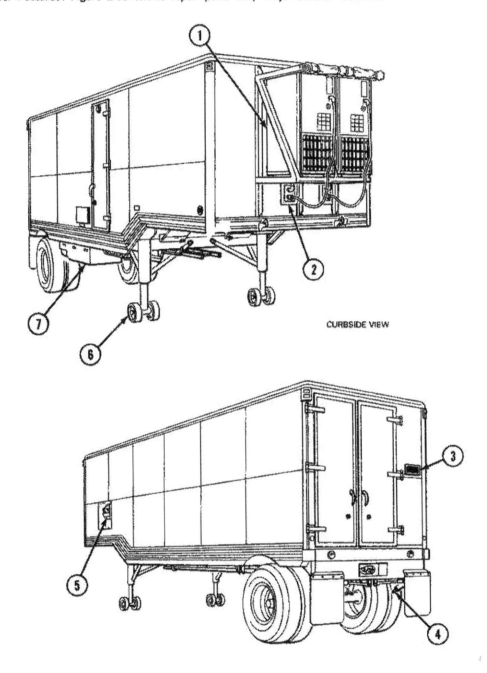

CURBSIDE VIEW

ROADSIDE VIEW

Figure 2-39. Repair Parts Shop Major Exterior Features

2-23. REPAIR PARTS SHOP – CONTINUED.

1   AIR CONDITIONER. Two units, each capable of 18,000 Btu cooling, 12,000 Btu heating, and ventilation.

2   CONNECTOR PANEL. Two connector panels supply 3-phase 208 V ac 50/60 Hz power to air conditioners.

3   AIR VENT. When opened from inside, provides filtered outside air.

4   LEVELING JACK AND FOOTPAD. Two leveling jacks and footpads support and level rear end of shop.

5   CABLE ENTRY PANEL. Used to connect power and ground cables and field telephone lines.

6   LANDING GEAR. Supports forward end of shop when disconnected from towing tractor.

7   STORAGE BOX. Storage for power and ground cables and for ground rods.

2-23. REPAIR PARTS SHOP -- CONTINUED.

b. *Interior Features.* Figure 2-40 shows repair parts shop major interior features.

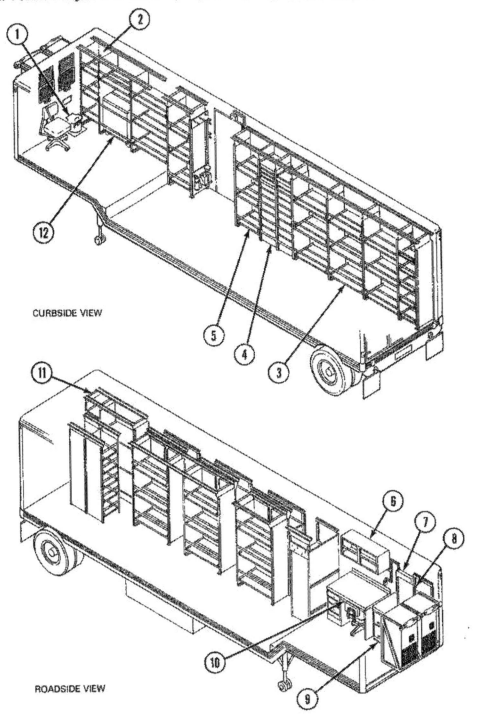

CURBSIDE VIEW

ROADSIDE VIEW

Figure 2-40. Repair Parts Shop Major Interior Features

2-23. REPAIR PARTS SHOP – CONTINUED.

1 DETONATOR/INITIATOR STORAGE BOX. Lockable storage area for missile detonators and initiators.

2 STORAGE RACK. Ten-shelf, general purpose open storage rack with webbing straps for storage and transportation of contents.

3 STORAGE RACK. One five-shelf and one nine-shelf general purpose open storage rack with webbing straps for storage and transportation of contents.

4 STORAGE DRAWER. Sixteen large drawers, six small drawers, and one small shelf for small parts and assemblies.

5 STORAGE RACK. Three-shelf, general purpose open storage rack with webbing straps for storage and transportation of contents.

6 BOOKRACK. Contains shelves with webbing straps for storage and transportation of contents.

7 ELECTRICAL DISTRIBUTION BOX. Provides control and distribution of circuit breaker-protected ac power to lights, convenience outlets, oven, and air conditioners in shop. Also contains connections for field telephones.

8 STORAGE RACK. Three-shelf, general purpose open storage rack with webbing straps for storage and transportation of contents.

9 MOBILE FILE CABINET. General purpose, two-drawer four-wheel file cabinet with locking wheels and worktable.

10 DESK. Used for general office procedures.

11 STORAGE RACK. Five three-shelf and five five-shelf, general purpose, open storage racks with webbing straps for storage and transportation of contents.

12 STORAGE AREA. Storage area with webbing straps for storage and transportation of mobile file cabinet.

2-24. REFERENCE SCENE GENERATION FACILITY (RSGF).

The RSGF is a mobile, self-contained facility used for making target reference scene tape cartridges. The tape cartridges are loaded into the PLC. The RSGF equipment is housed in an S-280 facilitized electrical equipment shelter mounted on an M928 cargo truck. The truck tows a trailer-mounted 30 kW generator that provides power for the RSGF. For more information on the RSGF, see TM 9-1430-388-12.

    a. *Exterior Features*. Figure 2-41 shows RSGF major exterior features.

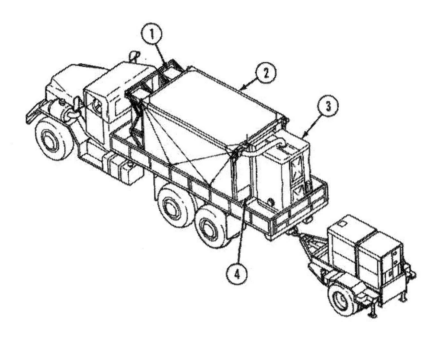

Figure 2-41. RSGF Major Exterior Features

2-24. REFERENCE SCENE GENERATION FACILITY (RSGF) – CONTINUED.

1   AIR CONDITIONER. Two units, each capable of 18,000 Btu cooling, 12,000 Btu heating, and ventilation.

2   FACILITIZED SHELTER. Contains RSGF equipment.

3   PE. Provides 5-minute air bath to remove CB contaminants from personnel before entering RSGF.

4   POWER ENTRY PANEL. Connects power, ground, and telephone cables to RSGF.

2-24. REFERENCE SCENE GENERATION FACILITY (RSGF) – CONTINUED.

b. *Interior Features.* Figure 2-42 shows RSGF major interior features.

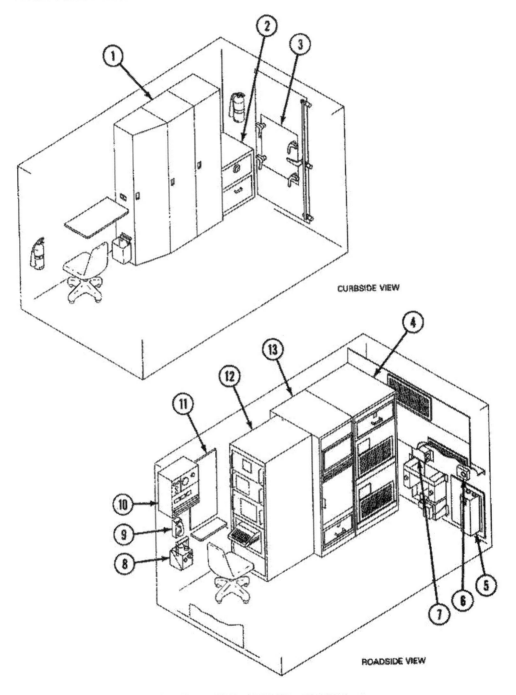

Figure 2-42. RSGF Major Interior Features

2-24. REFERENCE SCENE GENERATION FACILITY (RSGF) – CONTINUED.

1 STORAGE CABINET. Three cabinets provide storage for technical manuals, tools, and some disc packs.

2 SAFE. Provides storage for system disc pack, target cartridges, operator's log, and other material.

3 EMERGENCY PANEL. Provides emergency exit.

4 CABINET 1A3. Memory cabinet contains two disc drives and storage drawer.

5 CONVERTER. Changes input voltage to 208 V 400 Hz for operation of CB equipment.

6 CHEMICAL AGENT ALARM M42. Alerts personnel when detector triggers alarm unit.

7 CONTROL MODULE. Provides single point from which CB system status is monitored.

8 CHEMICAL AGENT DETECTOR M43. Triggers alarm unit when chemical agents are detected.

9 TELEPHONE TA-312/PT. Used for telephone communications.

10 POWER DISTRIBUTION PANEL. Provides circuit breaker-protected ac and dc power to components.

11 MESSAGE BOARD. Provides space for crew notes.

12 CABINET 1A1. Operator's console contains cartridge tape unit, printer, cathode ray tube (CRT) display, alphanumeric display, keyboard, and 28 V dc power supply.

13 CABINET 1A2. Computer cabinet contains computer control panel, three computer chassis, and storage drawer.

2-25. SYSTEM COMPONENTS TEST STATION (SCTS).

The SCTS is a mobile, self-contained facility for DS/GS-level testing and maintenance of PIi system components and assemblies. The SCTS equipment is housed in a modified M1006 van towed by an M931 tractor. Power is applied to the SCTS by a 60 kW generator mounted on a RAPU. For more information on the SCTS, see TM 9-4935-393-14-1.

2-25. SYSTEM COMPONENTS TEST STATION (SCTS) – CONTINUED.

a. *Exterior Features.* Figure 2-43 shows SCTS major exterior features.

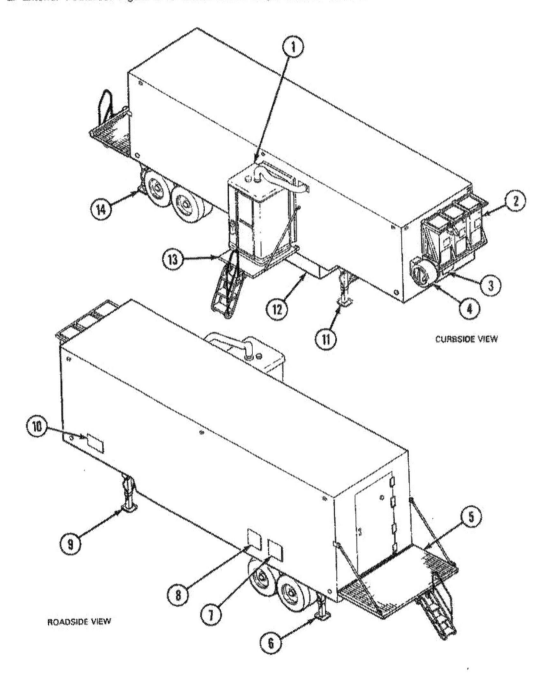

Figure 2-43. SCTS Major Exterior Features

2-25. SYSTEM COMPONENTS TEST STATION (SCTS) – CONTINUED.

1 PE. Provides 5-minute air bath to remove CB contaminants from personnel before entering SCTS.

2 AIR CONDITIONER. Three units, each capable of 18,000 Btu cooling, 12,000 Btu heating, and ventilation.

3 FORWARD CABLE ENTRY PANEL. Contains EME-filtered cable connections for air conditioners and CB filter unit.

4 CB FILTER UNIT. Contains gas and particulate filters with blower to force filtered air into SCTS.

5 REAR WORK PLATFORM. Provides work area for large assemblies.

6 LEVELING JACK. Levels van during emplacement.

7 COMMUNICATION CABLE ENTRY PANEL. Contains EME-filtered cable and headset connections between inside and outside of van.

8 SIGNAL CABLE ENTRY PANEL. Contains EME-filtered cable connections between inside and outside of van.

9 LANDING GEAR. Supports front of van.

10 POWER CABLE ENTRY PANEL. Contains EME-filtered cable connections for power cables from RAPU.

11 LANDING GEAR. Supports front of van.

12 STORAGE CABINET. Storage location for PE, side work platform, handrails, and canvas bags containing camouflage support equipment.

13 SIDE PERSONNEL PLATFORM. Platform for PE.

14 LEVELING JACK. Levels van during emplacement.

2-25. SYSTEM COMPONENTS TEST STATION (SCTS) – CONTINUED.

b. *Interior Features.* Figure 2-44 shows SCTS major interior features.

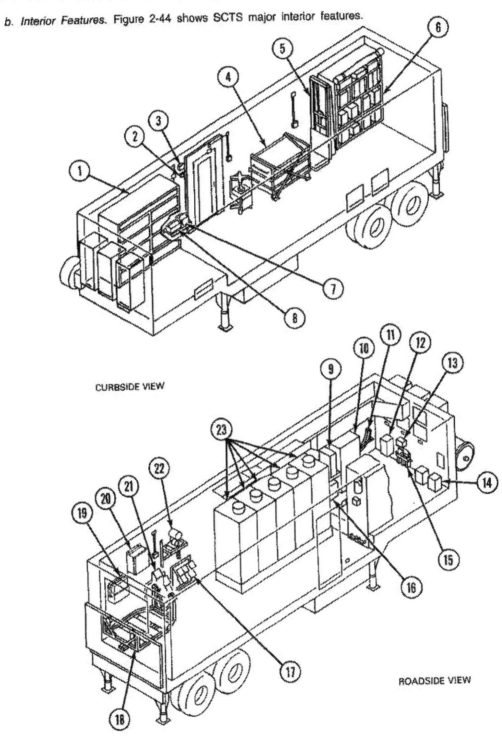

CURBSIDE VIEW

ROADSIDE VIEW

Figure 2-44. SCTS Major Interior Features

2-25. SYSTEM COMPONENTS TEST STATION (SCTS) – CONTINUED.

1   FORWARD STORAGE CABINET. Stores adapters used with automatic test set (ATS), cables, self-test jumper caps, and related items.

2   CHEMICAL AGENT ALARM M42. Alerts personnel by audible and visual alarms when chemical agents are detected.

3   PE BALST VALVE. Controls air flow to PE. Also provides EME-filtered cable connection to PE control module.

4   MOVABLE WORKTABLE. Used as work surface and storage for small items such as headsets and multimeter.

5   LIFT TRUCK. Used to lift and move heavy components and to remove IEU from G&C/A.

6   REAR STORAGE CABINET. Stores adapters used with ATS and test program disc packs.

7   CHEMICAL AGENT DETECTOR M43. Triggers chemical agent alarm when chemical agents are detected.

8   CHEMICAL AGENT DETECTOR POWER SUPPLY PS3. Provides power to operate chemical agent detector.

9   OSCILLOSCOPE. Used during manual testing of units under test (UUT's).

10 POWER DISTRIBUTION CABINET. Provides power distribution, EME filtering, overload protection, and voltage monitoring.

11 LSC COVERS. Protective covers for LSC used during in-van G&C/A tests.

12 CONVERTER. Converts 3-phase 208 V 60 Hz power to 3-phase 208 V 400 Hz for operation of CB equipment.

13 CB CONTROL MODULE. Contains controls and monitors necessary for operation of CB equipment.

14 HEATER. Two portable electric space heaters that provide additional heating where required.

15 FORWARD CABLE ENTRY PANEL. Contains EME filters and connectors for power distribution to air conditioners and CB filter unit.

16 28-VOLT POWER SUPPLY PS1. Provides 28 V dc to ATS for UUT testing.

17 SIGNAL CABLE ENTRY PANEL. Contains EME-filtered cable connections between inside and outside of van.

18 G&C/A DOLLY. Holds G&C/A during in-van testing.

19 DISTORTION MEASUREMENT SET. Used during UUT testing.

20 SELF-TEST CARD STORAGE CABINET. Stores digital adapter self-test cards.

21 COMMUNICATION CABLE ENTRY PANEL. Contains EME-filtered cable and headset connections between inside and outside of van.

22 G&C/A BLOWER. Provides forced air cooling for G&C/A during in-van testing.

23 ATS. Computer-controlled test equipment that tests PII assemblies.

2-26. REAR AREA POWER UNIT (RAPU).

The RAPU is a self-contained, trailer-mounted power station that is towed by an M925 cargo truck. The RAPU provides power for the SCTS, conditioned air for in-container testing of G&C/A sections, high-pressure air for missile controls and turbine testing, and storage for ground networks cables. Figure 2-45 shows RAPU major features. For more information on the RAPU, see TM 9-1450-394-14.

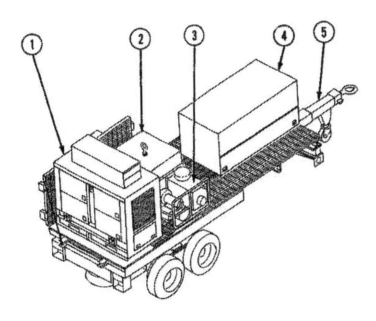

Figure 2-45. RAPU Major Features

1 60 KW GENERATOR. Provides 60 kW 3-phase 115 V ac 60 Hz power to SCTS, G&C/A conditioning assembly, and air servicer unit (ASU).
2 ASU. Contains compressor that charges four air receivers, providing reservoir of high-pressure air used for missile vane and nozzle tests.
3 G&C/A CONDITIONING ASSEMBLY. Provides heated or ambient air for G&C/A section in-container testing.
4 CABLE STORAGE CABINET. Provides storage for cables, ground rods, adapters, tools, and airhoses.
5 M796 TRAILER. Modified 4-ton, four-wheel trailer used for mounting components that make up RAPU.

2-27. 30 KW GENERATOR.

The 30 kW generator is a diesel-engine-driven power source. Its output is 3-phase 120/208 V ac 50/60 Hz power. The generator can be skid-mounted or trailer-mounted. The skid-mounted version is mounted on the EL tractor to supply power to the EL. The trailer-mounted version is mounted on a two-wheeled M200A1 trailer that is towed by either the PCC or RSGF to supply power to them. It is also towed to emplacement sites to provide power to the BCC, P&P shop, and repair parts shop. Figure 2-46 shows the 30 kW generator. For more information on the 30 kW generator, see TM 5-6115-465-12.

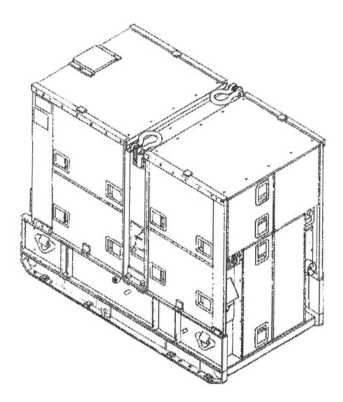

Figure 2-46. 30 kW Generator

2-28. TOOL KITS.

Tool kits used to maintain the PII system are:

a. *Contact Team Tool Kit.* This kit is used to perform DS/GS maintenance on mechanical and electrical assemblies. The kit is stored in contact team locations. For more information on the contact team tool kit, see SC 5180-92-CL-P24.

b. *Digital Repair Tool Kit.* This kit is used to perform DS/GS maintenance on electronic equipment. The kit is stored in the electrical repair shop and the SCTS. For more information on the digital repair tool kit, see SC 5180-92-CL-P23.

c. *Electrical Shop Tool Kit.* This kit is used to perform DS/GS maintenance on electrical equipment. The kit is stored in the electrical repair shop. For more information on the electrical shop tool kit, see SC 5180-92-CL-P22.

d. *Firing Site Tool Kit.* This kit is used to perform missile assembly and disassembly and organizational maintenance on firing battery equipment. The kit is stored with firing battery equipment at the discretion of the officer in charge (OIC). For more information on the firing site tool kit, see SC 5180-92-CL-P21.

2-28. TOOL KITS -- CONTINUED.

e. General Support Unit (GSU) Tool Kit. This kit contains specialized tools used to bend hydraulic or pneumatic tubes, to etch metal, etc. The kit is stored with rear area equipment at the discretion of the OIC. For more information on the GSU tool kit, see SC 5180-92-CL-P28.

f. Load Test Tool Kit. This kit is used to test slings and lifting devices. The kit is stored in the mechanical repair shop. For more information on the load test tool kit, see SC 5180-92-CL-P27.

g. Mechanical Shop Tool Kit. This kit is used to perform DS/GS maintenance on mechanical assemblies. The kit is stored in the mechanical repair shop. For more information on the mechanical shop tool kit, see SC 5180-92-CL-P26.

h. Mechanics and Structures Tool Kit. This kit is used to perform DS/GS maintenance on vehicle bodies, metal frames, etc. The kit is stored in the mechanical repair shop. For more information on the mechanics and structures tool kit, see SC 5180-92-CL-P25.

i. Special Tools and Equipment Tool Kit. This kit is used to perform special maintenance on ground handling equipment. The kit is stored with rear area equipment at the discretion of the OIC. For more information on the special tools and equipment tool kit, see SC 5180-92-CL-P29.

j. Storage Maintenance Tool Kit. This kit is used to perform organizational maintenance on equipment in storage. The kit is stored with firing battery equipment at the discretion of the OIC. For more information on the storage maintenance tool kit, see SC 5180-92-CL-P20.

2-29. CABLE SETS.

The forward area cable set and the rear area cable set are used to electrically connect units of the PII missile system (PCC to EL, RAPU to SCTS, etc.). For more information on the cable sets, see TM 9-1430-393-14.

a. Forward Area Cable Set. The forward area cable set connects forward area equipment. It is stored in the auxiliary vehicle. Figure 2-47 shows forward area cable set major features.

2-29. CABLE SETS – CONTINUED.

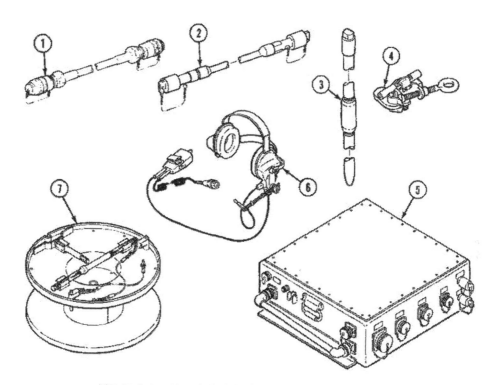

Figure 2-47. Forward Area Cable Set Major Features

1   CABLE ASSEMBLIES 60W43 AND 500W2 THROUGH 500W8. Electrically connect PII equipment.
2   CABLE ASSEMBLIES 60W101 AND 500W37. Five cable assemblies 60W101 and one cable assembly 500W37 used with ground rod sets and ground rod clamps to ground PII equipment.
3   GROUND ROD SET. Two sets used with cable assemblies 60W101 and ground rod clamps to ground PII equipment.
4   GROUND ROD CLAMP. Four clamps attach cable assemblies 60W101 to ground rods.
5   POWER DISTRIBUTION BOX (EL TRACTOR). Distributes power between 30 kW generator and EL.
6   HEADSET. Four headsets for firing platoon communications.
7   CABLE REEL. Stores cable assembly 500W6.

2-29. CABLE SETS – CONTINUED.

b. *Rear Area Cable Set*. The rear area cable set connects rear area equipment and is used to perform tests on assemblies. It is stored on the RAPU. Figure 2-48 shows rear area cable set major features.

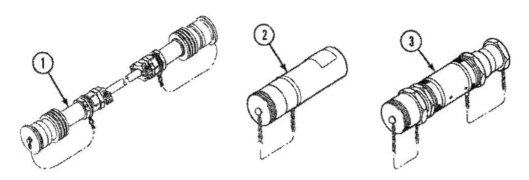

Figure 2-48. Rear Area Cable Set Major Features

1 CABLE ASSEMBLIES 60W43 AND 500W13 THROUGH 500W19. Electrically connect PII equipment.
2 DUMMY CONNECTORS 90A12 THROUGH 90A24. Used in self-tests to check PII system test equipment.
3 CABLE ADAPTERS 90A6 THROUGH 90A11, 90A25, AND 90A26. Used as extension cables for in-container test of missile.

2-30. WARHEAD FUNCTIONAL SIMULATOR (WFS).

The WFS is used to verify the PII missile before a tactical warhead is mated. The simulator consists of two cables and a dummy connector that provide electrical continuity and simulate a WS. Figure 2-49 shows WFS major features. For more information on the WFS, see TM 9-1430-393-14.

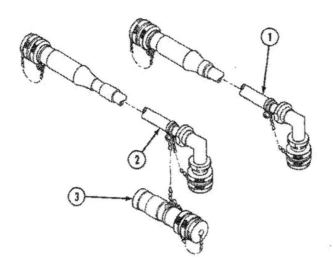

Figure 2-49. WFS Major Features

2-30. WARHEAD FUNCTIONAL SIMULATOR (WFS) -- CONTINUED.

1 CABLE ASSEMBLY 500W24. Used with cable assembly 500W25 to electrically connect RS and G&C/A.
2 CABLE ASSEMBLY 500W25. Used with cable assembly 500W24 to electrically connect RS and G&C/A.
3 DUMMY CONNECTOR 500W25P3. Used with cable assemblies 500W24 and 500W25 to simulate WS.

2-31. FACILITIZED ELECTRICAL EQUIPMENT SHELTER.

The facilitized electrical equipment shelter is a modified S-280 shelter that is used for the PCC and RSGF. It provides EME protection and portability of PCC and RSGF equipment. Major shelter items common to the PCC and RSGF are:

- Air conditioners
- Power entry and distribution panels
- Lighting system
- CB protection system.

For more information on the facilitized electrical equipment shelter, see TM 9-1425-391-14.

2-32. 60 KW GENERATOR.

The 60 kW generator is a diesel-engine-driven power source. Its output is 3-phase 120/208 V ac 50/60 Hz power. The generator can be skid-mounted or trailer-mounted. The skid-mounted version is mounted on the RAPU to supply power to the SCTS. The trailer-mounted version is mounted on a two-wheeled M200A1 trailer that is towed to emplacement sites to supply power to the electrical and mechanical repair shops. Figure 2-50 shows the 60 kW generator. For more information on the 60 kW generator, see TM 5-6115-545-12.

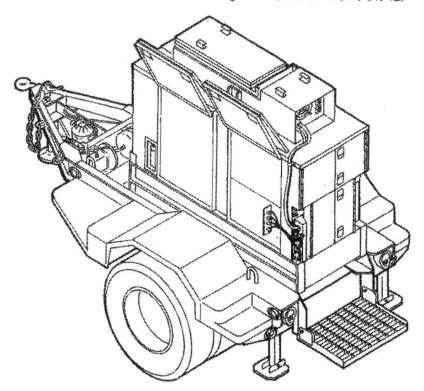

Figure 2-50. 60 kW Generator

2-33. CAS SITE POWER DISTRIBUTION EQUIPMENT.

CAS site power distribution equipment allows operation of the CAS site with commercial power. The equipment is used to distribute tactical and commercial power at the CAS site. Figure 2-51 shows CAS site power distribution equipment. For more information on CAS site power distribution equipment, see TM 9-1430-393-14.

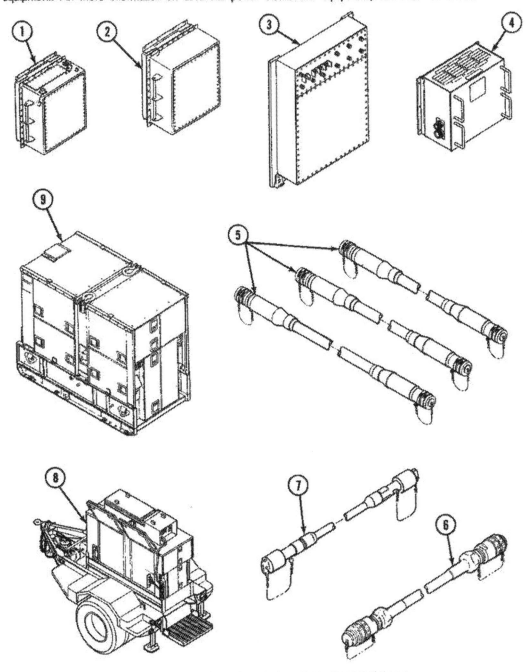

Figure 2-51. CAS Site Power Distribution Equipment

2-33. CAS SITE POWER DISTRIBUTION EQUIPMENT – CONTINUED.

1 FILTER BOX (CAS PAD). Distributes 3-phase ac power and 28 V dc power from power distribution box (CAS pad) to EL and system data recorder.

2 POWER DISTRIBUTION/FILTER BOX (CAS VAN). Distributes 3-phase ac power from 30 kW generator or commercial power source to CAS van.

3 POWER DISTRIBUTION BOX (CAS PAD). Distributes 3-phase ac power from 60 kW generator or commercial power source to pad filter assemblies. Distributes 28 V dc to power distribution/filter box (CAS van) and 60 kW generator battery.

4 28 V DC POWER SUPPLY. Provides 28 V dc to power distribution box (CAS pad).

5 CABLE ASSEMBLY 500W6. Nine cable assemblies electrically connect CAS van to EL GIEU's.

6 CABLE ASSEMBLY 500W2. Two cable assemblies electrically connect power distribution/filter box (CAS van) to 30 kW generator and CAS van.

7 CABLE ASSEMBLY 60W101. Two cable assemblies ground power distribution/filter box (CAS van) to 30 kW generator and CAS van.

8 60 KW GENERATOR. Three stationary 60 kW generators provide 3-phase ac power to three power distribution boxes (CAS pad).

9 30 KW GENERATOR. Trailer-mounted 30 kW generator provides 3-phase ac power to power distribution/filter box (CAS van).

2-34. SCTS FILTER ASSEMBLY.

The SCTS filter assembly allows operation of the SCTS with use of commercial power. Figure 2-52 shows SCTS filter assembly major features. For more information on the SCTS filter assembly, see TM 9-1430-393-14.

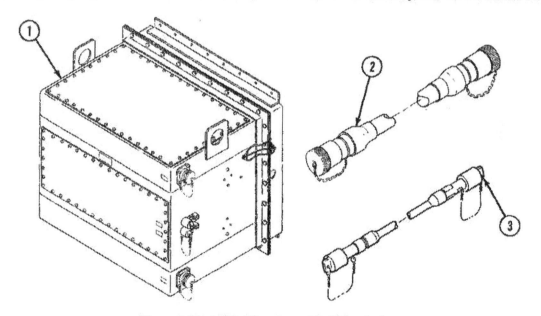

Figure 2-52. SCTS Filter Assembly Major Features

1 SCTS FILTER BOX. Distributes 3-phase ac power from commercial power source to RAPU power distribution box. Used instead of 60 kW generator to provide power to SCTS.

2 CABLE ASSEMBLY 500W11. Two cables electrically connect SCTS filter box to RAPU power distribution box.

3 CABLE ASSEMBLY 60W101. Grounds SCTS filter box to junction of existing cable assemblies 60W101 from SCTS and UUT container.

Section IV. TRAINING DEVICES

2-35. MODIFIED TACTICAL (MT) MISSILE SECTION TRAINERS.

MT missile section trainers are used for initial and follow-on training in support of missile assembly, missile section replacement and handling, component replacement, inspection, and preventive maintenance checks and services (PMCS). These trainers are similar to tactical missile sections in appearance, size, weight, center of gravity, and function. For training safety, ordnance items have been removed and the RS magnetron has been disabled. The G&C/A section battery is replaced with a battery simulator. Mating surfaces between sections have been modified to prevent accidental mating with tactical missile sections. In addition, inert training items are used. For more information on the MT missile section trainers, see TM 9-6920-387-14.

2-36. MISSILE ASSEMBLY SIMULATED COUNT (MASC) MISSILE SECTION TRAINERS.

MASC missile section trainers are used for initial and follow-on training in support of PII missile assembly, missile section replacement and handling, service upon receipt (except in-container tests by SCTS), ignition safe and arm checks, coding and recoding of ignition enable, radar section pressurization, system countdown, and malfunction isolation. Component parts in these trainers are similar to tactical components only in appearance, size, weight, center of gravity, and method of attachment. MASC missile section trainers have only the circuits required to respond as tactical sections during simulated countdown. Mating surfaces do not allow accidental mating with tactical sections. Software programs used with MASC missile section trainers provide simulated countdown and reference scene data. Simulated faults also can be inserted for training in fault isolation. For more information on MASC missile section trainers, see TM 9-6920-387-14.

2-37. FIRST AND SECOND STAGE COUNTDOWN TRAINERS.

First and second stage countdown trainers are used for initial and follow-on training in support of missile assembly, missile section replacement and handling, countdown operations, and UUT testing (except controls tests). These trainers are similar to tactical propulsion sections in appearance, size, weight, center of gravity, and method of attachment. The trainers do not contain propellant or ordnance items. The first stage trainer also contains a simulated S&A cover and nozzle and no fin control hardware. The trainers can be mated with tactical or trainer G&C/A sections. However, mating surfaces do not allow accidental mating with tactical propulsion sections. For more information on first and second stage countdown trainers, see TM 9-6920-387-14.

2-37.1. MISSILE SIMULATOR ES-1.

Missile simulator ES-1 is used to train selected personnel in the overall operation of the EL with or without missile section trainers or a tactical missile. The simulator allows the student to learn countdown procedures for the GIEU, RLCU, and EL operations. Figure 2-53 shows missile simulator ES-1 major features. For more information on missile simulator ES-1, see TM 9-6920-387-14.

2-37. MISSILE SIMULATOR ES-1 – CONTINUED.

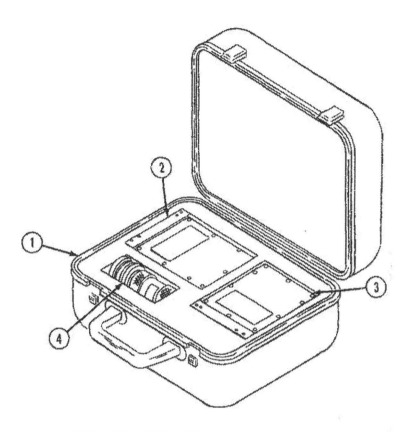

Figure 2-53. Missile Simulator ES-1 Major Features

1 CARRYING CASE. Provides protection and means of transport for simulator components.
2 TRAINING TARGET TAPE CARTRIDGE. Provides simulated reference scene information to LCA of GIEU.
3 TRAINING COUNTDOWN TAPE CARTRIDGE. Provides training program information to LCA of GIEU.
4 DUMMY CONNECTOR PLUG 20A1P1. Required for ES-1 training operations without training missile sections. Used to electrically simulate tactical missile.

2-38. GIEU SIMULATOR.

The GIEU simulator provides hands-on training and experience in the operation and function of the GIEU and RLCU. The GIEU simulator allows students to perform EL countdown and maintenance operations in a classroom environment. Using GIEU controls, the instructor can preset countdown conditions and enter simulated holds and malfunctions. GIEU simulator controls and indicators function as they would in a tactical countdown. The GIEU simulator also electronically simulates the following EL functions:

* Azimuth clamp
* Hydraulics
* Leveling jacks and sensor
* Pallet rotation.

2-38. GIEU SIMULATOR – CONTINUED.

The GIEU simulator can operate with a tactical PCC, PCC trainer, or RLCU. Students operating the GIEU simulator can communicate with PCC, PCC trainer, or RLCU operators by headset intercom. The GIEU simulator is mounted on casters and can be moved by two persons. Figure 2-54 shows GIEU simulator major features. For more information on the GIEU simulator, see TM 9-6920-389-14.

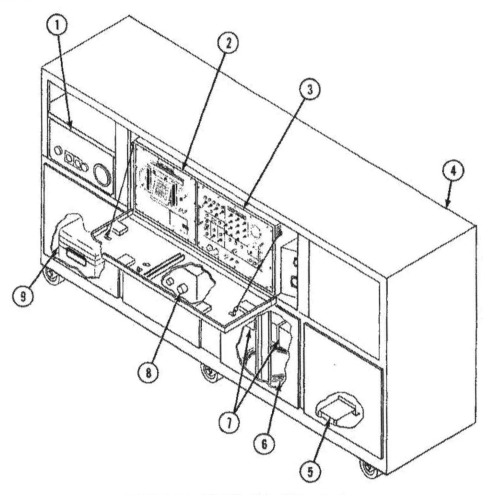

Figure 2-54. GIEU Simulator Major Features

1   HYDRAULIC CONTROL PANEL. Contains controls and indicators for simulated hydraulic operations.
2   LCA. Tactical assembly that performs same functions as LCA on EL.
3   PCA. Tactical assembly that performs same functions as PCA on EL.
4   CONSOLE. Contains all GIEU simulator components. Mounted on casters for positioning.
5   SIMULATOR BOARD. Electronically simulates azimuth clamp, hydraulic, rotating pallet, and leveling jacks operations.
6   CABLES 500W27 AND 500W28. Power cable 500W27 and RLCU cable 500W28.
7   TAPE CARTRIDGES. Two tape cartridges: one training countdown tape cartridge and one training target reference scene tape cartridge.
8   POWER SUPPLY PS1. Tactical EL power supply. Provides 28 V dc to GIEU assembly.
9   RLCU. Tactical assembly that performs same functions as RLCU in PCC.

2-39. PCC TRAINER.

The PCC trainer provides experience for PII personnel in:

- Data safeguards
- Countdown and status display for up to three missiles
- Alphanumeric display of event sequence for all missiles
- RLCU deployment and operation
- Launch time data entry, automatic computation, and display of launch window data
- Printout of hard copy for time-referenced countdown.

A separate instructor's console interfaces with the trainer equipment. The instructor uses this console to:

- Monitor the student's progress
- Control previously inserted simulated malfunctions
- Activate microprocessor-controlled simulation of missile/RLCU functions
- Control application of power to the trainer.

The instructor's console is also compatible with a tactical PCC. Headset communications link the trainer and the instructor's console. Figure 2-55 shows PCC trainer major features. For more information on the PCC trainer, see TM 9-6920-392-14.

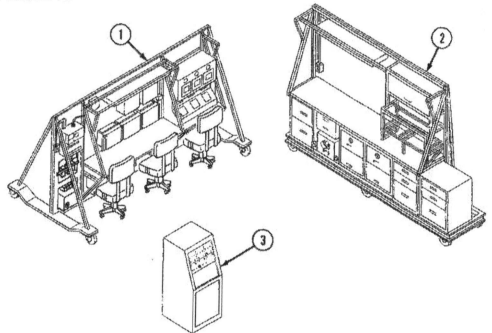

Figure 2-55. PCC Trainer Major Features

1 ROADSIDE CONSOLE. Contains equipment located on roadside of tactical PCC, including power distribution panel, power entry panel, signal entry panel, status display panel, launch window panel, ILA, and three RLCU's.

2 CURBSIDE CONSOLE. Contains equipment located on curbside of tactical PCC, including storage drawers, communications equipment rack, security cabinets, and removable safe.

3 INSTRUCTOR CONSOLE. Contains equipment required to control application of power to trainer, activate microprocessor-controlled simulation of missile/RLCU functions, control previously inserted simulated malfunctions, and monitor student progress.

2-40. RSGF TRAINER.

The RSGF trainer is used during classroom instruction in RSGF operation. The trainer consists of tactical RSGF equipment removed from the shelter and mounted on castered frames. Simulated malfunctions can be inserted by loading special software into the RSGF computer. Figure 2-56 shows RSGF trainer major features. For more information on the RSGF trainer, see TM 9-6920-388-14.

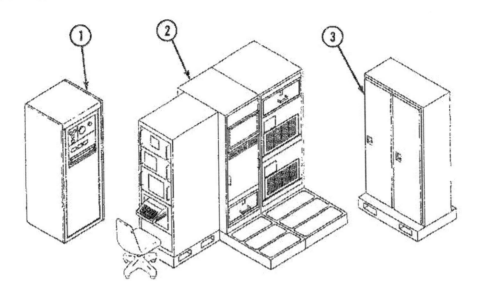

Figure 2-56. RSGF Trainer Major Features

  1 POWER DISTRIBUTION CABINET. Consists of tactical RSGF power distribution panel.

  2 ROADSIDE EQUIPMENT. Consists of tactical RSGF equipment mounted on roadside of shelter, including operator's console, computer cabinet, and memory cabinet.

  3 CURBSIDE EQUIPMENT. Consists of storage cabinets for technical manuals, tools, and disc packs.

2-41. SCTS TRAINER.

The SCTS trainer provides training in operation and maintenance of SCTS equipment and UUT's. The SCTS trainer enables students to perform SCTS self-test and UUT test procedures in a classroom environment. Special software and a self-test adapter make it possible to insert faults that cause simulated ATS analog and digital component failures. UUT tests can be run with faulty tactical assemblies to aid in fault recognition and response. Figure 2-57 shows SCTS trainer major features. For more information on the SCTS trainer, see TM 9-6920-393-14.

2-41. SCTS TRAINER – CONTINUED.

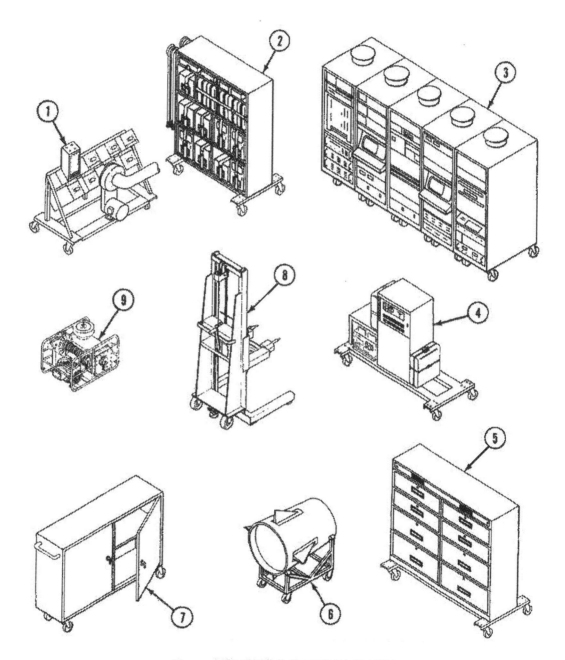

Figure 2-57. SCTS Trainer Major Features

1 SIGNAL ENTRY PANEL. Contains components of tactical SCTS signal cable entry and communication cable entry panels. Also holds G&C/A blower assembly.

2 REAR STORAGE CABINET. Storage for UUT adapters, self-test cards, cables, disc packs, LSC covers, distortion measurement set, and DTU test fixture.

2-41. SCTS TRAINER – CONTINUED.

3 ATS. Same as tactical ATS.

4 POWER DISTRIBUTION ASSEMBLY. Consists of SCTS power distribution cabinet and 28 V dc power supply. Storage for oscilloscope.

5 FORWARD STORAGE CABINET. Storage for same items stored in tactical SCTS forward storage cabinet, such as adapters, cables, and self-test jumper caps. Storage for trainer digital self-test adapter.

6 G&C/A DOLLY. Same as G&C/A dolly used in tactical SCTS. Holds G&C/A during test.

7 MOVABLE WORKTABLE. Used as work surface and storage for small items such as headsets and multimeter.

8 LIFT TRUCK. Same as lift truck used in tactical SCTS. Lifts and moves heavy components.

9 G&C/A CONDITIONING ASSEMBLY. Same as G&C/A conditioning assembly stored on RAPU. Provides heated or ambient air for G&C/A section in-container testing.

CHAPTER 3
OPERATIONAL DEPLOYMENT

For information on command and control, firing roles, and missions, see FM 6-11 and FM 6-12.

CHAPTER 4
MAINTENANCE AND LOGISTICS

Section I. MAINTENANCE

4-1. MAINTENANCE CONCEPT.

a. *General.* The PII maintenance concept places most of the maintenance responsibility outside of the firing batteries. This allows the firing batteries to concentrate on fulfilling their primary firing missions. The following factors determine how the maintenance will be accomplished:

- Status of the Army
- Status of the firing battery
- Location of the battery
- Availability of parts and tools
- Availability of personnel.

b. *Peacetime Conditions.* During peacetime, the battery on CAS has the highest priority. The CAS battery performs only maintenance essential to continual readiness. All other preventive maintenance is deferred until the battery is released from this role. Corrective maintenance for the CAS battery normally consists of replacing a faulty assembly. The replacement can come from an operational readiness float (ORF) or the release battery. DS/GS maintenance contact teams from the FSC provide additional support. Batteries not on CAS are maintained in accordance with their role:

- Pre-CAS -- next to assume CAS; has next highest priority
- Maneuver – just released from CAS; next priority after pre-CAS
- Release – maximum maintenance.

4-2. CATEGORIES OF MAINTENANCE.

a. *General.* The PII maintenance concept is carried out by three categories of maintenance:
- Unit (Organizational) maintenance
- Intermediate (DS/GS) maintenance
- Depot maintenance.

Each category responds to technical problems with personnel technical training and maintenance facilities specific to that category.

4-2. CATEGORIES OF MAINTENANCE – CONTINUED.

b. *Unit (Organizational) Maintenance.* Unit maintenance is provided by the unit using the equipment. The maintenance is carried out at firing sites and in garrison by personnel in the firing units. Troubleshooting normally consists of operator observations of displays, indicators, meters, gages, and general equipment condition. Corrective maintenance is generally limited to replacement of major assemblies (for example, missile sections, LCA, RLCU) or easily replaced items such as indicators. Faulty assemblies are turned over to DS/GS maintenance personnel for repair.

c. *Intermediate (DS/GS) Maintenance.* Intermediate maintenance is provided by two support battalion organizations. These are the FSC's and the Maintenance and Supply (M&S) company. An FSC is assigned to each field artillery battalion. The FSC can support organizational maintenance facilities or function on its own. It uses the electrical and mechanical repair shops, SCTS, and RAPU to troubleshoot and repair assemblies. These assemblies can be forward or rear area equipment. The M&S company uses the same facilities as the FSC's to maintain the ORF's. There is one M&S company for each brigade. Items beyond the repair capability of the intermediate-forward (DS) maintenance organization are returned to the theatre class V intermediate-rear (GS) maintenance organization for repair (first stage and second stage propulsion sections) or are returned to depot for repair.

d. *Depot Maintenance.* Depot maintenance uses tools and test equipment similar to those used by the manufacturer. These enable a depot facility to perform repairs up to and including complete overhaul if required. Depot-repaired items are returned to the supply system.

Section II. LOGISTICS

4-3. STOCKPILE-TO-FIRING BATTERY SEQUENCE.

The materiel goes from the manufacturer to a depot for distribution. During predeployment verification and tests, the prime contractor is responsible for resupply of PII equipment. After deployment, normal supply channels are used for resupply. Figure 4-1 shows the stockpile-to-firing battery sequence.

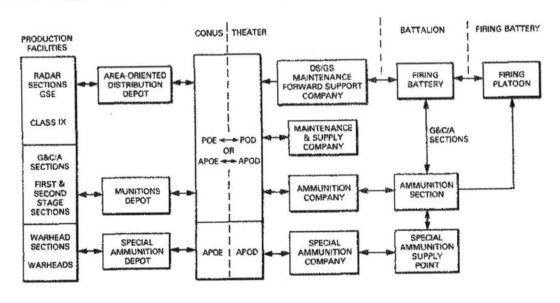

Figure 4-1. Stockpile-to-Firing Battery Sequence

4-4. RESUPPLY REQUISITION FLOW.

Requisitions for missiles and equipment originate within the battalion. Requisitions from the firing battery are routed through the headquarters (HQ) and service battery. The requisitions are then routed to the theater stock control agency. Theater depot stocks of PII items are established before system deployment. Theater stocks of missiles and ORF's of GSE are stored as directed by the theater commander. CONUS stocks are maintained at authorized stations and depots. Figure 4-2 shows the resupply requisition flow.

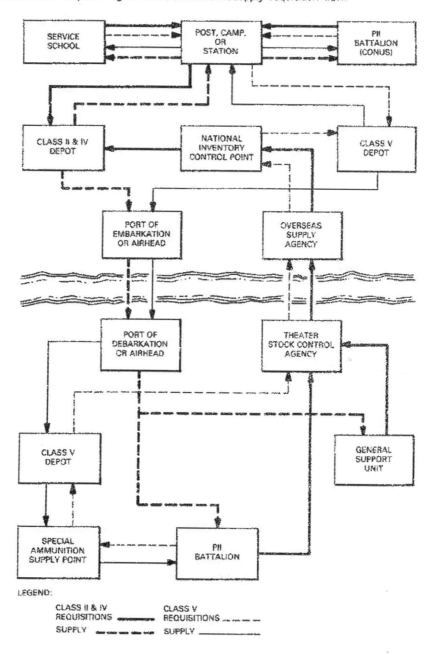

Figure 4-2. Resupply Requisition Flow

4-5. MISSILE RESUPPLY FLOW.

The battalion obtains a resupply of missile sections from the special ammunition supply point. The HQ and service battery ammunition platoon transports the missile sections. Upon receipt, the missile sections are inspected and tested. An inventory is also conducted. The missile sections are then placed in storage until requested by a firing battery. Figure 4-3 shows the missile resupply flow.

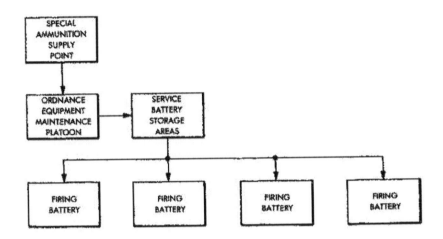

Figure 4-3. Missile Resupply Flow

5-2. SHIPPING AND STORAGE CONTAINERS DATA.

Dimensions and weight for each container are listed in table 5-2.

Table 5-2. Container Data

Container	Length in. (m)	Width in. (m)	Height in. (m)	Empty Weight Lb (kg)
First or second stage aft skirt	70.5 (1.8)	53.01 (1.4)	58.92 (1.5)	1,195 (542.0)
First stage	190 (4.8)	72 (1.8)	68.59 (1.7)	5,290 (2,399.5)
Second stage	129 (3.3)	72 (1.8)	68.59 (1.7)	4,350 (1,973.1)
G&C/A section	97 (2.5)	72 (1.8)	68.59 (1.7)	2,700 (1,224.7)
Warhead section	97.75 (2.5)	47.44 (1.2)	51.75 (1.3)	1,505 (683.3)
Radar section	78 (2.0)	37 (0.9)	45.75 (1.2)	1,265 (573.8)

5-3. GROUND HANDLING EQUIPMENT DATA.

Aft skirt holding fixture
Manufacturer's part number 13088001
Weight .... 250 lb (113.4 kg)

EL tractor accessory kit
Manufacturer's part number 11500285
Weight (less 30 kW generator and power distribution box) .... 275 lb (124.7 kg)

First stage hoisting beam
Manufacturer's part number 11500280
Weight .... 490 lb (222.3 kg)
Lifting capacity .... 12,100 lb (5,488.5 kg)

Four-leg sling assembly
Manufacturer's part number 11027819
Lifting capacity .... 9,200 lb (4,154.5 kg)

Four-leg sling assembly
Manufacturer's part number 11500994
Weight (without leg extensions) .... 230 lb (104.3 kg)
Weight (with leg extensions) .... 433 lb (196.4 kg)
Lifting capacity .... 15,000 lb (6,803.9 kg)

Guided missile section cradle
Manufacturer's part number 13088002
Weight .... 2,172 lb (985.3 kg)

Radar section cover set
Manufacturer's part number 13090182
Weight .... 77.1 lb (35.0 kg)

Second stage hoisting beam
Manufacturer's part number 11500281
Weight .... 380 lb (172.4 kg)
Lifting capacity .... 12,100 lb (5,488.5 kg)

Shelter tiedown kit
Manufacturer's part number 13088003
Weight .... 166 lb (75.3 kg)

5-3. GROUND HANDLING EQUIPMENT DATA – CONTINUED.

Two-leg sling
    Manufacturer's part number 11500991
    Weight .... 83 lb (37.6 kg)
    Lifting capacity .... 15,000 lb (6,803.9 kg)

Universal sling
    Manufacturer's part number 11500282
    Weight .... 45.9 lb (20.8 kg)
    Lifting capacity .... 800 lb (326.9 kg)

30 kW generator set accessory kit (EL tractor)
    Manufacturer's part number 11500458
    Weight .... 20.4 lb (9.3 kg)

30 kW generator set accessory kit (trailer)
    Manufacturer's part number 13086024
    Weight .... 20.4 lb (9.3 kg)

60 kW generator set accessory kit (RAPU)
    Manufacturer's part number 11500378
    Weight .... 12 lb (5.4 kg)

60 kW generator set accessory kit (trailer)
    Manufacturer's part number 13155466
    Weight .... 12 lb (5.4 kg)

5-4. EL DATA.

EL overall dimensions are shown in figure 5-1. Additional equipment data follows the figure.

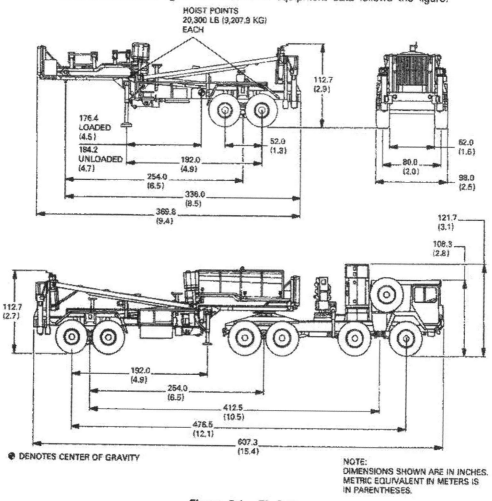

Figure 5-1. EL Data

Weights

EL with missile ready for road march	39,601 lb (17,962.7 kg)
EL less missile ready for road march	23,165 lb (10,507.5 kg)
Front jack assembly	350 lb (158.8 kg)
LCA	125 lb (56.7 kg)
PCA	85 lb (38.6 kg)
Missile power supply	87 lb (39.5 kg)
EL power supply	87 lb (39.5 kg)
Rear jack assembly	370 lb (167.8 kg)

Wheel assembly data

Wheel size	14 x 22.5
Tire size	18 x 22.5
Ply rating	16 ply
Operating air pressure	55 psig (379 kPa)
Wheel and tire assembly weight	270 lb (122.5 kg)

5-4. EL DATA – CONTINUED.

Hydraulic tank capacity (filled to upper line on sight glass) .... 33 gal (124.9 l)

Power requirements

Voltage .... 24 V dc, 120/208 V ac, 3-phase, 4-wire

Frequency .... 50/60 Hz ±5 percent

Power .... 30 kW, power factor 80 percent

Pneumatic requirements

For hydraulic accumulator (precharge) .... Two stage 1600 to 1800 psig (11,032 to 12,411 kPa)

EL emplacement limitations

Ground slope .... 6°

5-5. FACILITIZED ELECTRICAL EQUIPMENT SHELTER DATA.

Weight .... 3,938 lb (1,786.2 kg)

Length .... 149 in. (3.8 m)

Width .... 87 in. (2.2 m)

Height .... 88 in. (2.2 m)

5-6. PCC DATA.

Weight and dimensions

Truck mounted:

Weight .... 34,486 lb (15,642.6 kg)

Length .... 450 in. (10.3 m)

Width .... 97 in. (2.5 m)

Height .... 150 in. (3.8 m)

PCC only:

Weight .... 6,675 lb (3,027.7 kg)

Length .... 188 in. (4.8 m)

Width .... 87 in. (2.2 m)

Height .... 94 in. (2.4 m)

Power requirements

Ac:

Voltage .... 120/208 V, 3-phase, 4-wire

Frequency .... 50/60 Hz ±5 percent

Power .... 30 kW, power factor 80 percent

Dc:

Backup power, 24 V provided by truck battery.

5-7. BCC DATA.

BCC outer dimensions are shown in figure 5-2. Additional equipment data follows the figure.

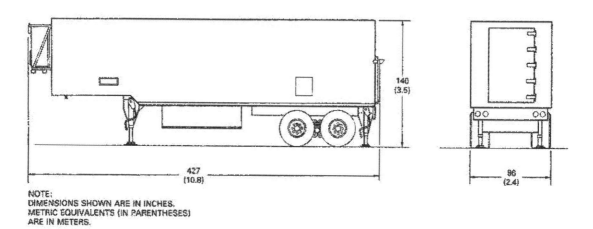

NOTE:
DIMENSIONS SHOWN ARE IN INCHES.
METRIC EQUIVALENTS (IN PARENTHESES)
ARE IN METERS.

Figure 5-2. BCC Data

BCC

Power requirements:
  Voltage .... 120/208 V 30 kW
  Line frequency .... 50/60 Hz
  Phase .... 3-phase, 4-wire wye connected
Power at convenience outlets .... 120 V 50/60 Hz
EME attenuation:
  200 ± 10 kHz .... −50 dB
  21 ± 0.5 MHz .... −60 dB
Maximum operational altitude .... 8,000 ft (2,440 m)

Semitrailer van M1006

Weight (loaded) .... 38,500 lb (17,463.3 kg)
Tire inflation:
  Highway .... 70 psi (482.7 kPa)
  Cross-country .... 45 psi (310.3 kPa)
  Sand, mud, snow .... 45 psi (310.3 kPa)
Dimensions:
  Length .... 427 in. (10.8 m)
  Width .... 96 in. (2.4 m)
  Height .... 140 in. (3.5 m)

Tractor M931

Weight:
  Fuel tanks full .... 22,173 lb (10,057.5 kg)
  Fuel tanks empty .... 21,353 lb (9,685.6 kg)
Length .... 273 in. (6.9 m)
Width .... 98 in. (2.5 m)
Height .... 112 in. (2.8 m)

5-7. BCC DATA – CONTINUED.

Towing speeds

Highway and improved roads .... 50 mph (80 kph)

Unimproved roads .... 20 mph (32 kph)

Rough terrain and cross-country .... 10 mph (16 kph)

Air conditioner

Model F18T-2S

National stock number 4120-01-114-2471

Manufacturer's part number 13225E8000

Technical manuals .... TM 5-4120-371-14, TM 5-4120-371-24P

Power requirements (per unit):

    Voltage .... 208 V

    Line frequency .... 50/60 Hz

    Phase 3-phase

Current (maximum) .... 20.8 A

Cooling capacity (per unit) .... 18,000 Btu/hr (maximum)

Heating capacity (per unit) .... 12,000 Btu/hr (maximum)

Weight (per unit) .... 280 lb (127.0 kg)

Dimensions (per unit):

    Width .... 17.3 in. (0.4 m)

    Depth .... 20 in. (0.5 m)

    Height .... 46.5 in. (1.2 m)

Fire extinguisher

Charging chemical .... Monobromotrifluoromethane

Charge .... 2-3/4 lb (1.3 kg)

5-8. CAS VAN DATA.

CAS van outer dimensions are shown in figure 5-3. Additional equipment data follows the figure.

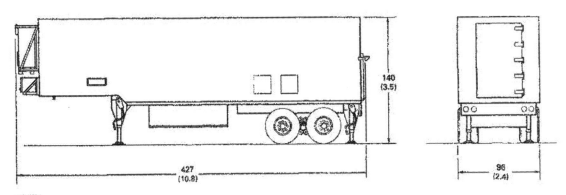

NOTE:
DIMENSIONS SHOWN ARE IN INCHES.
METRIC EQUIVALENTS (IN PARENTHESES)
ARE IN METERS.

Figure 5-3. CAS Van Data

5-8. CAS VAN DATA – CONTINUED.

CAS van

    Power requirements:
        Voltage .... 120/208 V 30 kW
        Line frequency .... 50/60 Hz
        Phase .... 3-phase, 4-wire wye connected
    Power at convenience outlets .... 120 V 50/60 Hz
    EME attenuation:
        200 ± 10 kHz .... −50 dB
        21 ± 0.5 MHz .... −60 dB
    Maximum operational altitude .... 8,000 ft (2,440 m)

Semitrailer van M1006

    Weight (loaded) .... Approximately 40,800 lb (18,506.6 kg)
    Tire inflation:
        Highway .... 70 psi (482.7 kPa)
        Cross-country .... 45 psi (310.3 kPa)
        Sand, mud, snow .... 45 psi (310.3 kPa)
    Dimensions:
        Length .... 427 in. (10.8 m)
        Width .... 96 in. (2.4 m)
        Height .... 140 in. (3.5 m)

Tractor M931

    Weight:
        Fuel tanks full .... 22,173 lb (10,057.5 kg)
        Fuel tanks empty .... 21,353 lb (9,685.6 kg)
    Length .... 273 in. (6.9 m)
    Width .... 98 in. (2.5 m)
    Height .... 112 in. (2.8 m)

Towing speeds

    Highway and improved roads .... 50 mph (80 kph)
    Unimproved roads .... 20 mph (32 kph)
    Rough terrain and cross-country .... 10 mph (16 kph)

Air conditioner

    Model F18T-2S
    National stock number 4120-01-114-2471
    Manufacturer's part number 13225E8000
    Technical manuals .... TM 5-4120-371-14, TM 5-4120-371-24P
    Power requirements (per unit):
        Voltage .... 208 V
        Line frequency .... 50/60 Hz
        Phase 3-phase
    Current (maximum) .... 20.8 A
    Cooling capacity (per unit) .... 18,000 Btu/hr (maximum)
    Heating capacity (per unit) .... 12,000 Btu/hr (maximum)
    Weight (per unit) .... 280 lb (127.0 kg)

5-8. CAS VAN DATA – CONTINUED.

Dimensions (per unit):
Width .... 17.3 in. (0.4 m)
Depth .... 20 in. (0.5 m)
Height .... 46.5 in. (1.2 m)

Fire extinguisher
Charging chemical .... Monobromotrifluoromethane
Charge .... 2-3/4 lb (1.3 kg)

5-9. RSGF DATA.

Weight and dimensions
Truck mounted:
Weight .... 38,771 lb (17,586.2 kg)
Length .... 450 in. (11.4 m)
Width .... 97 in. (2.5 m)
Height .... 150 in. (3.8 m)

RSGF only:
Weight .... 7,937 lb (3,600.2 kg)
Length .... 188 in. (4.8 m)
Width .... 87 in. (2.2 m)
Height .... 94 in. (2.4 m)

Power requirements
Voltage .... 120/208 V ac, 3-phase, 4-wire
Frequency .... 50/60 Hz ±5 percent
Power .... 30 kW, power factor 80 percent

5-10. SCTS DATA.

SCTS outer dimensions are shown in figure 5-4. Additional equipment data follows the figure.

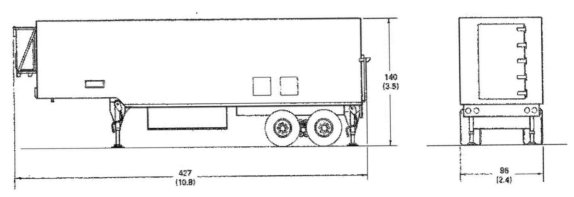

NOTE:
DIMENSIONS SHOWN ARE IN INCHES.
METRIC EQUIVALENTS (IN PARENTHESES)
ARE IN METERS.

Figure 5-4. SCTS Data

5-10. SCTS DATA – CONTINUED.

SCTS

Power requirements:
Voltage .... 120/208 V 32.5 kW
Line frequency .... 50/60 Hz
Phase .... 3-phase, 4-wire wye connected
Power at convenience outlets .... 120 V 50/60 Hz
EME attenuation:
200 ± 10 kHz .... –50 dB
21 ± 0.5 MHz .... –60 dB
Maximum operational altitude .... 8,000 ft (2,440 m)

Semitrailer van M1006

Weight:
Loaded .... 39,035 lb (17,706.0 kg)
Without camouflage nets (6) .... 38,237 lb (17,344.0 kg)
Length .... 427 in. (10.8 m)
Width .... 96 in. (2.4 m)
Height .... 140 in. (3.5 m)

Tractor M931

Weight:
Fuel tanks full .... 22,173 lb (10,057.5 kg)
Fuel tanks empty .... 21,353 lb (9,685.6 kg)
Length .... 273 in. (6.9 m)
Width .... 98 in. (2.5 m)
Height .... 112 in. (2.8 m)

Towing speeds

Highway and improved roads .... 50 mph (80 kph)
Unimproved roads .... 30 mph (48 kph)
Rough terrain and cross-country .... 5 mph (8 kph)

Air conditioner

Model F18T-2S
National stock number 4120-01-114-2471
Manufacturer's part number 13225E8000
Technical manuals .... TM 5-4120-371-14, TM 5-4120-371-24P

Power requirements (per unit):
Voltage .... 208 V
Line frequency .... 50/60 Hz
Phase 3-phase
Current (maximum) .... 20.8 A
Cooling capacity (per unit) .... 18,000 Btu/hr (maximum)
Heating capacity (per unit) .... 12,000 Btu/hr (maximum)
Weight (per unit) .... 280 lb (127.0 kg)
Dimensions (per unit):
Width .... 17.3 in. (0.4 m)
Depth .... 20 in. (0.5 m)
Height .... 46.5 in. (1.2 m)

5-10. SCTS DATA – CONTINUED.

Chemical agent alarm system
Power requirements:
Voltage .... 120/208 V
Line frequency .... 50/60 Hz
Phase 3-phase
Power consumption .... 1.5 kW

Fire extinguisher
Charging chemical .... Monobromotrifluoromethane
Charge .... 2-3/4 lb (1.3 kg)

ATS
Manufacturer's part number 13085950
Power requirements:
Voltage .... 120/208 V +5/–10 percent
Phase 3-phase
Line frequency .... 50/60 Hz ± 5 percent
Power consumption .... 15.6 kW
Operational environment:
Temperature .... 70°F ± 10°F (21°C ± 5°C)
Relative humidity .... 80 percent (maximum)

5-11. ELECTRICAL REPAIR SHOP DATA.

Electrical repair shop outer dimensions are shown in figure 5-5. Additional equipment data follows the figure.

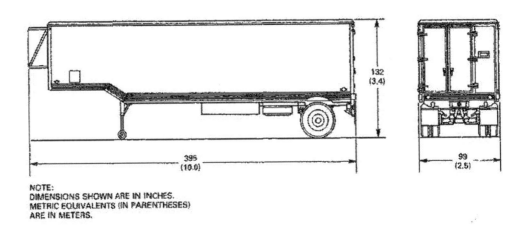

NOTE:
DIMENSIONS SHOWN ARE IN INCHES.
METRIC EQUIVALENTS (IN PARENTHESES)
ARE IN METERS.

Figure 5-5. Electrical Repair Shop Data

5-11. ELECTRICAL REPAIR SHOP DATA – CONTINUED.

Air conditioner

Model F18T-2S
National stock number 4120-01-114-2471
Manufacturer's part number 13225E8000
Technical manuals .... TM 5-4120-371-14, TM 5-4120-371-24P
Power requirements (per unit):
    Voltage .... 208 V
    Line frequency .... 50/60 Hz
    Phase 3-phase
Current (maximum) .... 20.8 A
Cooling capacity (per unit) .... 18,000 Btu/hr (maximum)
Heating capacity (per unit) .... 12,000 Btu/hr (maximum)
Weight (per unit) .... 280 lb (127.0 kg)
Dimensions (per unit):
    Width .... 17.3 in. (0.4 m)
    Depth .... 20 in. (0.5 m)
    Height .... 46.5 in. (1.2 m)

Electrical repair shop

Power requirements:
    Voltage .... 120/208 V 60 kW
    Line frequency .... 50/60 Hz
    Phase .... 3-phase, 4-wire wye connected
Power at convenience outlets .... 120 V 50/60 Hz
Maximum operational altitude .... 8,000 ft (2.4 km)

Fire extinguisher

Charging chemical .... Monobromotrifluoromethane
Charge .... 2-3/4 lb (1.3 kg)

Semitrailer van M373A2 modified

Weight (loaded) .... 15,106 lb (6,852.0 kg)
Tire inflation:
    Highway .... 50 psi (345 kPa)
    Cross-country .... 35 psi (241 kPa)
    Sand, mud, snow .... 15 psi (103 kPa)
Dimensions:
    Length .... 395 in. (10.0 m)
    Width .... 99 in. (2.5 m)
    Height .... 132 in. (3.4 m)

Towing speeds

Highway and improved roads .... 50 mph (80 kph)
Unimproved roads .... 20 mph (32 kph)
Rough terrain and cross-country .... 10 mph (16 kph)

Tractor M931

Weight .... 22,173 lb (10,057.5 kg)
Length .... 273 in. (6.9 m)
Width .... 98 in. (2.5 m)
Height .... 112 in. (2.8 m)

5-12. MECHANICAL REPAIR SHOP DATA.

Mechanical repair shop outer dimensions are shown in figure 5-6. Additional equipment data follows the figure.

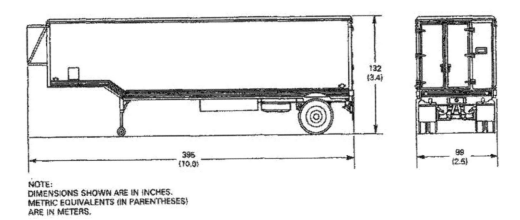

NOTE:
DIMENSIONS SHOWN ARE IN INCHES.
METRIC EQUIVALENTS (IN PARENTHESES)
ARE IN METERS.

Figure 5-6. Mechanical Repair Shop Data

Air conditioner
    Model F18T-2S
    National stock number 4120-01-114-2471
    Manufacturer's part number 13225E8000
    Technical manuals .... TM 5-4120-371-14, TM 5-4120-371-24P

    Power requirements (per unit):
        Voltage .... 208 V
        Line frequency .... 50/60 Hz
        Phase 3-phase
    Current (maximum) .... 20.8 A
    Cooling capacity (per unit) .... 18,000 Btu/hr (maximum)
    Heating capacity (per unit) .... 12,000 Btu/hr (maximum)
    Weight (per unit) .... 280 lb (127.0 kg)

    Dimensions (per unit):
        Width .... 17.3 in. (0.4 m)
        Depth .... 20 in. (0.5 m)
        Height .... 46.5 in. (1.2 m)

Fire extinguisher
    Charging chemical .... Monobromotrifluoromethane
    Charge .... 2-3/4 lb (1.3 kg)

Mechanical repair shop
    Power requirements:
        Voltage .... 120/208 V 60 kW
        Line frequency .... 50/60 Hz
        Phase .... 3-phase, 4-wire wye connected
    Power at convenience outlets .... 120 V 50/60 Hz
    Maximum operational altitude .... 8,000 ft (2.4 km)

5-12. MECHANICAL REPAIR SHOP DATA -- CONTINUED.

Semitrailer van M373A2 modified

Weight (loaded) .... 15,900 lbs (7,212.1 kg)

Tire inflation:

    Highway .... 50 psi (345 kPa)

    Cross-country .... 35 psi (241 kPa)

    Sand, mud, snow .... 15 psi (103 kPa)

Dimensions:

    Length .... 395 in. (10.0 m)

    Width .... 99 in. (2.5 m)

    Height .... 132 in. (3.4 m)

Towing speeds

Highway and improved roads .... 50 mph (80 kph)

Unimproved roads .... 20 mph (32 kph)

Rough terrain and cross-country .... 10 mph (16 kph)

Tractor M931

Weight .... 22,173 lb (10,057.5 kg)

Length .... 273 in. (6.9 m)

Width .... 98 in. (2.5 m)

Height .... 112 in. (2.8 m)

5-13. P&P SHOP DATA.

P&P shop outer dimensions are shown in figure 5-7. Additional equipment data follows the figure.

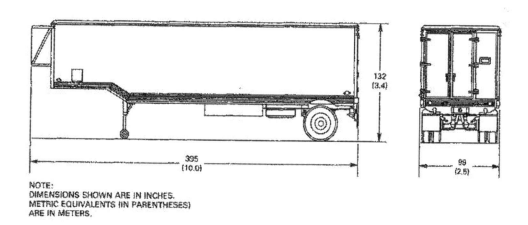

NOTE:
DIMENSIONS SHOWN ARE IN INCHES.
METRIC EQUIVALENTS (IN PARENTHESES)
ARE IN METERS.

Figure 5-7. P&P Shop Data

5-13. P&P SHOP DATA – CONTINUED.

Air conditioner

Model F18T-2S
National stock number 4120-01-114-2471
Manufacturer's part number 13225E8000
Technical manuals .... TM 5-4120-371-14, TM 5-4120-371-24P

Power requirements (per unit):
  Voltage .... 208 V
  Line frequency .... 50/60 Hz
  Phase 3-phase
Current (maximum) .... 20.8 A
Cooling capacity (per unit) .... 18,000 Btu/hr (maximum)
Heating capacity (per unit) .... 12,000 Btu/hr (maximum)
Weight (per unit) .... 280 lb (127.0 kg)

Dimensions (per unit):
  Width .... 17.3 in. (0.4 m)
  Depth .... 20 in. (0.5 m)
  Height .... 46.5 in. (1.2 m)

Fire extinguisher

Charging chemical .... Monobromotrifluoromethane
Charge .... 2-3/4 lb (1.3 kg)

P&P shop

Power requirements:
  Voltage .... 120/208 V 30 kW
  Line frequency .... 50/60 Hz
  Phase .... 3-phase, 4-wire wye connected
Power at convenience outlets .... 120 V 50/60 Hz
Maximum operational altitude .... 8,000 ft (2.4 km)

Semitrailer van M373A2 modified

Weight (loaded) .... 15,632 lb (7,090.6 kg)
Tire inflation:
  Highway .... 50 psi (345 kPa)
  Cross-country .... 35 psi (241 kPa)
  Sand, mud, snow .... 15 psi (103 kPa)
Dimensions:
  Length .... 395 in. (10.0 m)
  Width .... 99 in. (2.5 m)
  Height .... 132 in. (3.4 m)

Tractor M931

Weight .... 22,173 lb (10,057.5 kg)
Length .... 273 in. (6.9 m)
Width .... 98 in. (2.5 m)
Height .... 112 in. (2.8 m)

Towing speeds

Highway and improved roads .... 50 mph (80 kph)
Unimproved roads .... 20 mph (32 kph)
Rough terrain and cross-country .... 10 mph (16 kph)

5-14. REPAIR PARTS SHOP DATA.

Repair parts shop outer dimensions are shown in figure 5-8. Additional equipment data follows the figure.

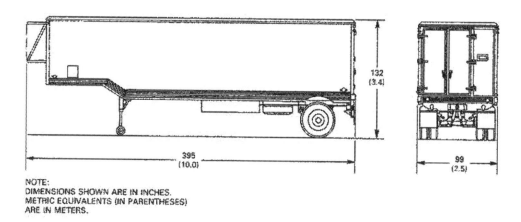

NOTE:
DIMENSIONS SHOWN ARE IN INCHES.
METRIC EQUIVALENTS (IN PARENTHESES)
ARE IN METERS.

Figure 5-8. Repair Parts Shop Data

Air conditioner

Model F18T-2S
National stock number 4120-01-114-2471
Manufacturer's part number 13225E8000
Technical manuals .... TM 5-4120-371-14, TM 5-4120-371-24P
Power requirements (per unit):
    Voltage .... 208 V
    Line frequency .... 50/60 Hz
    Phase 3-phase
Current (maximum) .... 20.8 A
Cooling capacity (per unit) .... 18,000 Btu/hr (maximum)
Heating capacity (per unit) .... 12,000 Btu/hr (maximum)
Weight (per unit) .... 280 lb (127.0 kg)
Dimensions (per unit):
    Width .... 17.3 in. (0.4 m)
    Depth .... 20 in. (0.5 m)
    Height .... 46.5 in. (1.2 m)

Fire extinguisher

Charging chemical .... Monobromotrifluoromethane
Charge .... 2-3/4 lb (1.3 kg)

Repair parts shop

Power requirements:
    Voltage .... 120/208 V 30 kW
    Line frequency .... 50/60 Hz
    Phase .... 3-phase, 4-wire wye connected
Power at convenience outlets .... 120 V 50/60 Hz
Maximum operational altitude .... 8,000 ft (2.4 km)

5-14. REPAIR PARTS SHOP DATA – CONTINUED.

Semitrailer van M373A2 modified
- Weight (loaded) .... 13,727 lb (6,226.5 kg)
- Tire inflation:
  - Highway .... 50 psi (345 kPa)
  - Cross-country .... 35 psi (241 kPa)
  - Sand, mud, snow .... 15 psi (103 kPa)
- Dimensions:
  - Length .... 395 in. (10.0 m)
  - Width .... 99 in. (2.5 m)
  - Height .... 132 in. (3.4 m)

Tractor M931
- Weight .... 22,173 lb (10,057.5 kg)
- Length .... 273 in. (6.9 m)
- Width .... 96 in. (2.5 m)
- Height .... 112 in. (2.8 m)

Towing speeds
- Highway and improved roads .... 50 mph (80 kph)
- Unimproved roads .... 20 mph (32 kph)
- Rough terrain and cross-country .... 10 mph (16 kph)

5-15. RAPU DATA.

RAPU
- Manufacturer's part number 11500306
- Weight .... 12,820 lb (5,815.0 kg)
- Power output:
  - Voltage .... 120/208 V 60 kW
  - Line frequency .... 60 Hz
  - Phase .... 3-phase, 4 wye
- High-pressure air output .... 1300 ± 150 psig
- EME attenuation:
  - 200 ± 10 kHz .... −50 dB
  - 21 ± 0.5 MHz .... −60 dB

Trailer
- Model M796
- Manufacturer's part number 11500350
- Weight .... 4,820 lb (2,186.4 kg)
- Maximum towed speed:
  - Highway .... 50 mph (80.5 kph)
  - Cross-country .... 10 mph (16.1 kph)
- Tire inflation:
  - Highway .... 45 lb (310.3 kPa)
  - Cross-country .... 25 lb (172.4 kPa)
  - Sand, mud, snow .... 25 lb (172.4 kPa)
- Dimensions:
  - Length (maximum extended) .... 258.5 in. (6.6 m)
  - Length (retracted) .... 210.5 in. (5.4 m)
  - Width .... 92 in. (2.3 m)
  - Height .... 42.5 in. (1.1 m)

5-15. RAPU DATA -- CONTINUED.

ASU

Model	MM-000100
Manufacturer's part number	11500304
High-pressure air storage capacity	320 cu ft at 3,000 psig
Power requirements:	
Voltage	120/208 V ac, +28 V dc
Line frequency	50/60 Hz
Phase	3-phase, 4 wye
Weight	600 lb (272.2 kg)
Dimensions:	
Width	36 in. (0.9 m)
Height	34 in. (0.9 m)
Length	42 in. (1.1 m)
Technical manuals	TM 9-1450-394-14, TM 9-1450-394-24P

G&C/A conditioning assembly

Manufacturer's part number	11500376
Weight	224 lb (101.6 kg)
Power requirements:	
Voltage	208 V ac
Line frequency	50/60 Hz
Phase	3-phase, 4 wye

5-16. TRANSPORTATION VEHICLES DATA.

This paragraph contains tables 5-3 through 5-6, which list the following vehicle data:

- Weights and dimensions
- Performance
- Tire pressure
- Capacities.

Table 5-3. Transportation Vehicles Weights and Dimensions

Vehicle	Weight (empty)	Height	Length	Width	Ground clearance	Wheel base
M818	20,995 lb[1] (9,523.2 kg)	116 in. (3.0 m)	280 in.[2] (7.1 m)	97.8 in. (2.5 m)	Variable[3] ---	167 in. (4.2 m)
M871	15,630 lb (7,089.6 kg)	103 in. (2.6 m)	358 in. (9.1 m)	96 in. (2.4 m)	18 in. (0.5 m)	---
M925	24,080 lb (10,922.5 kg)	116 in. (3.0 m)	329.1 in. (8.4 m)	98 in. (2.5 m)	11.5 in. (0.3 m)	179 in. (4.6 m)
M928	27,811 lb (12,615.0 kg)	116 in. (3.0 m)	404.9 in. (10.3 m)	98 in. (2.5 m)	11.5 in. (0.3 m)	215 in. (5.5 m)
M931	22,173 lb (10,057.5 kg)	111.8 in. (2.8 m)	272.6 in. (6.9 m)	98 in. (2.5 m)	11.5 in. (0.3 m)	167 in. (4.2 m)
M932	22,841 lb (10,360.5 kg)	111.8 in. (2.8 m)	286 in. (7.3 m)	98 in. (2.5 m)	11.5 in. (0.3 m)	167 in. (4.2 m)
M983	38,660 lb (17,535.9 kg)	112 in. (2.8 m)	351 in. (8.9 m)	96 in. (2.4 m)	24 in. (0.6 m)	181 in. (4.6 m)

5-16. TRANSPORTATION VEHICLES DATA -- CONTINUED.

Table 5-3. Transportation Vehicles Weights and Dimensions -- Continued

Vehicle	Weight (empty)	Height	Length	Width	Ground clearance	Wheel base
M984	41,574 lb (18,857.7 kg)	112 in. (2.8 m)	377 in. (9.6 m)	96 in. (2.4 m)	24 in. (0.6 m)	210 in. (5.3 m)
M1001	42,333 lb (19,201.9 kg)	122 in. (3.1 m)	337 in. (8.6 m)	99 in. (2.5 m)	18 in. (0.5 m)	244 in. (6.2 m)
M1002	41,901 lb (19,006.0 kg)	112 in. (2.8 m)	354 in. (9.0 m)	99 in. (2.5 m)	18 in. (0.5 m)	281 in. (7.1 m)

[1] 705 pounds (319.8 kg) less without front winch.

[2] 14 inches (0.4 m) less without front winch.

[3] Ground clearance is determined by type of trailer being towed.

Table 5-4. Vehicle Performance Data

Performance	M818	M925 and M928	M931	M932	M983	M984	M1001 and M1002
Speed (maximum)	54 mph (86.9 kph)	52 mph (83.7 kph)	52 mph (83.7 kph)	52 mph (83.7 kph)	57 mph (91.7 kph)	57 mph (91.7 kph)	57.4 mph (92.4 kph)
Front winch Capacity	20,000 lb (9,071.8 kg)	20,000 lb (9,071.8 kg)	None	20,000 lb (9,071.8 kg)	20,000 lb (9,071.8 kg)	20,000 lb (9,071.8 kg)	20,000 lb (9,071.8 kg)
Cable length	200 ft (61 m)	200 ft (61 m)	None	200 ft (61 m)	200 ft (61 m)	200 ft (61 m)	148 ft (45 m)
Recommended towed load, pintle On highway	30,000 lb (13,607.8 kg)	30,000 lb (13,607.8 kg)	30,000 lb (13,607.8 kg)	30,000 lb (13,607.8 kg)	20,000 lb (9,071.8 kg)	20,000 lb (9,071.8 kg)	11,085 lb (5,028.1 kg)
Off highway	15,000 lb (6,803.9 kg)	15,000 lb (6,803.9 kg)	15,000 lb (6,803.9 kg)	15,000 lb (6,803.9 kg)	20,000 lb (9,071.8 kg)	20,000 lb (9,071.8 kg)	11,085 lb (5,028.1 kg)
Payload (maximum) On highway	25,000 lb (11,339.8 kg)	20,000 lb (9,071.8 kg)	55,000 lb (24,947.6 kg)	55,000 lb (24,947.6 kg)	62,000 lb (28,122.7 kg)	52,000 lb (23,586.8 kg)	85,000 lb (38,555.4 kg)
Off highway	15,000 lb (6,803.9 kg)	10,000 lb (4,535.9 kg)	37,500 lb (17,009.7 kg)	37,500 lb (17,009.7 kg)	62,000 lb (28,122.7 kg)	52,000 lb (23,586.8 kg)	85,000 lb (38,555.4 kg)

Table 5-5. Vehicle Tire Pressure Data

Vehicle	Highway	Cross-Country	Mud, Snow, Sand
M818 Front	80 psig (552 kPa)	60 psig (414 kPa)	25 psig (172 kPa)
Rear	50 psig (345 kPa)	30 psig (207 kPa)	----
M871	75 psig (518 kPa)	75 psig (518 kPa)	---

5-16. TRANSPORTATION VEHICLES DATA – CONTINUED.

Table 5-5. Vehicle Tire Pressure Data – Continued

Vehicle	Highway	Cross-Country	Mud, Snow, Sand
M925, M928, M931, and M932			
Front	80 psig (552 kPa)	60 psig (414 kPa)	25 psig (172 kPa)
Rear	50 psig (345 kPa)	30 psig (207 kPa)	25 psig (172 kPa)
M983			
Front	60 psig (414 kPa)	35 psig (241 kPa)	20 psig (138 kPa)
Rear	70 psig (483 kPa)	40 psig (276 kPa)	30 psig (207 kPa)
M984			
Front	60 psig (414 kPa)	35 psig (241 kPa)	20 psig (138 kPa)
Rear	90 psig (620 kPa)	90 psig (620 kPa)	90 psig (620 kPa)
M1001			
Front	50 psig (345 kPa)	50 psig (345 kPa)	----
Rear	50 psig (345 kPa)	50 psig (345 kPa)	----
M1002			
Front	50 psig (345 kPa)	50 psig (345 kPa)	----
Rear	90 psig (620 kPa)	90 psig (620 kPa)	----

Table 5-6. Vehicle Capacities

Location	M818	M925 and M928	M931	M932	M983 and M984	M1001 and M1002
Cooling system	32 qt (30.3 l)	47 qt (44.5 l)	47 qt (44.5 l)	47 qt (44.5 l)	80 qt (75.7 l)	23.6 qt (22.3 l)
Crankcase (less oil filter)	28 qt (26.5 l)	23 qt (21.7 l)	23 qt (21.7 l)	23 qt (21.7 l)	22 qt (20.8 l)	26 qt (24.6 l)
Oil filter	2 qt (1.9 l)	4 qt (3.8 l)	4 qt (3.8 l)	4 qt (3.8 l)	2 qt (1.9 l)	5.7 qt (5.4 l)
Differentials						
1st front axle	12 qt (11.3 l)	12 qt (11.3 l)	12 qt (11.3 l)	12 qt (11.3 l)	18 qt (17.0 l)	5.5 qt (5.2 l)
2nd front axle	---	---	---	---	19.5 qt (18.4 l)	7.5 qt (7.1 l)
1st rear axle	12 qt (11.3 l)	12 qt (11.3 l)	12 qt (11.3 l)	12 qt (11.3 l)	19 qt (18.0 l)	13 qt (12.3 l)
2nd rear axle	12 qt (11.3 l)	12 qt (11.3 l)	12 qt (11.3 l)	12 qt (11.3 l)	18 qt (17.0 l)	11.5 qt (10.9 l)

5-16. TRANSPORTATION VEHICLES DATA – CONTINUED.

Table 5-6. Vehicle Capacities – Continued

Location	M818	M925 and M928	M931	M932	M983 and M984	M1001 and M1002
Transmission	11 qt* (10.4 l)	19 qt (18.0 l)	17 qt (16.0 l)	19 qt (18.0 l)	31 qt (29.3 l)	33.8 qt (31.9 l)
Fuel tank	110 gal (416.0 l)	81 gal (306.0 l)	116 gal (439.0 l)	116 gal (439.0 l)	154 gal (583.0 l)	105 gal (397.0 l)
Winch Front	1.3 qt (1.2 l)	1.3 qt (1.2 l)	---	1.3 qt (1.2 l)	---	1.05 qt (1.0 l)
Rear	---	---	---	---	---	1.05 qt (1.0 l)

* Capacity is 9 quarts without a power takeoff (PTO) gearbox.

CHAPTER 5
SYSTEM DATA

5-1. MISSILE DATA.

Dimensions and weight for each missile section and for the entire missile are listed in table 5-1.

Table 5-1. Missile Data

Missile Sections	Length In. (m)	Diameter In. (m) Max.	Min.	Weight Lb (kg)
First stage	144.74 (3.7)	40 (1.0)	—	9,156 (4,153.1)
Second stage	97.3 (2.5)	40 (1.0)	—	5,802 (2,631.7)
G&C/A section	61.51 (1.6)	40 (1.0)	27.75 (0.7)	669 (303.5)
Warhead section	64.25 (1.6)	27.7 (0.7)	20 (0.5)	591 (268.1)
Radar section	49.75 (1.3)	20 (0.5)	0.0	233 (105.7)
Missile	417.55 (10.6)	40 (1.0)	0.0	16,451 (7,462.0)

APPENDIX A
REFERENCES

See TM 9-1425-386-L, List of Applicable Publications (LOAP) for PERSHING II Field Artillery Missile System.

INDEX

INDEX — Continued

Made in the USA
Charleston, SC
17 December 2013